2015
万达商业规划
持有类物业

WANDA COMMERCIAL PLANNING 2015
PROPERTIES FOR HOLDING

万达商业规划研究院 主编

中国建筑工业出版社

EDITORIAL BOARD MEMBERS
编委会成员

主编单位
万达商业规划研究院

规划总指导
王健林

执行编委
赖建燕 叶宇峰 朱其玮 冯腾飞 孙培宇 方伟 侯卫华
杨旭 马红 刘冰

参编人员
沈文忠 王群华 曹春 范珑 黄引达 毛晓虎 闫红伟
罗沁 王玉龙 李浩 袁志浩 刘江 兰勇 葛宁 刘佩
张振宇 高振江 李斌 曹国峰 章宇峰 黄勇 蒲峰
朱欢 王雪松 徐立军 屈娜 刘安 万志斌 张宁 王朝忠
秦鹏华 王权 石路也 赵青扬 郭扬 王治天 李小强
王文广 张宝鹏 董明海 陈海亮 王宇石 陆峰 孙佳宁
李彬 张堃

耿大治 宋锦华 赵陨 黄路 张德志 王昉 孙辉 赵剑利

吴绿野 张涛 都晖 蓝毅 吴迪 党恩 孙海龙 沈余
陈杰 张鹏翔 虞朋 吕鲲 王吉 刘洋 郭晨光 朱迪
刘锋 方文奇 张堃 李涛 赵洪斌 刘志业 冯董 熊厚
黄涛 罗贤君 张琳 徐小莉 程欢 张洋 孟祥宾 董根泉
王静 李捷 关发扬

周澄 邓金坷 李江涛 钟光辉 陈勇 张剑锋 周德
李易 闫颇 唐杰 任意刚 张争 谷建芳 李舒仪 李昕
齐宗新 杨艳坤 冯志红 胡存珊 段堃 庞博

校对
马红 徐立军 张涛

英文校对
陈海亮 刘春 方芳 高建航 袁喆

CHIEF EDITORIAL UNIT
Wanda Commercial Planning & Research Institute

GENERAL PLANNING DIRECTOR
Wang Jianlin

EXECUTIVE EDITORIAL BOARD MEMBERS
Lai Jianyan, Ye Yufeng, Zhu Qiwei, Feng Tengfei, Sun Peiyu, Fang Wei, Hou Weihua, Yang Xu, Ma Hong, Liu Bing

PARTICIPANTS
Shen Wenzhong, Wang Qunhua, Cao Chun, Fan Long, Huang Yinda, Mao Xiaohu, Yan Hongwei, Luo Qin, Wang Yulong, Li Hao, Yuan Zhihao, Liu Jiang, Lan Yong, Ge Ning, Liu Pei, Zhang Zhenyu, Gao Zhenjiang, Li Bin, Cao Guofeng, Zhang Yufeng, Huang Yong, Pu Feng, Zhu Huan, Wang Xuesong, Xu Lijun, Qu Na, Liu An, Wan Zhibin, Zhang Ning, Wang Chaozhong, Qin Penghua, Wang Quan, Shi Luye, Zhao Qingyang, Guo Yang, Wang Zhitian, Li Xiaoqiang, Wang Wenguang, Zhang Baopeng, Dong Minghai, Chen Hailiang, Wang Yushi, Lu Feng, Sun Jianing, Li Bin, Zhang Kun

Geng Dazhi, Song Jinhua, Zhao Yun, Huang Lu, Zhang Dezhi, Wang Fang, Sun Hui, Zhao Jianli

Wu Lvye, Zhang Tao, Du Hui, Lan Yi, Wu Di, Dang En, Sun Hailong, Shen Yu, Chen Jie, Zhang Pengxiang, Yu Peng, Lv Kun, Wang Ji, Liu Yang, Guo Chenguang, Zhu Di, Liu Feng, Fang Wenqi, Zhang Kun, Li Tao, Zhao Hongbin, Liu Zhiye, Feng Dong, Xiong Hou, Huang Tao, Luo Xianjun, Zhang Lin, Xu Xiaoli, Cheng Huan, Zhang Yang, Meng Xiangbin, Dong Genquan, Wang Jing, Li Jie, Guan Fayang

Zhou Cheng, Deng Jinke, Li Jiangtao, Zhong Guanghui, Chen Yong, Zhang Jianfeng, Zhou De, Li Yi, Yan Po, Tang Jie, Ren Yigang, Zhang Zheng, Gu Jianfang, Li Shuyi, Li Xin, Qi Zongxin, Yang Yankun, Feng Zhihong, Hu Cunshan, Duan Kun, Pang Bo

PROOFREADERS
Ma Hong, Xu Lijun, Zhang Tao

ENGLISH VERSION PROOFREADERS
Chen Hailiang, Liu Chun, Fang Fang, Gao Jianhang, Yuan Zhe

CONTENTS
目录

A FOREWORD 010
序言

SUCCESS OF CORPORATE TRANSFORMATION DAWNING 012
企业转型初见成效

WANDA COMMERCIAL PLANNING 2015 014
万达商业规划 2015

B WANDA XISHUANGBANNA INTERNATIONAL RESORT 018
万达西双版纳国际度假区

01 WANDA XISHUANGBANNA INTERNATIONAL RESORT PLANNING 020
万达西双版纳国际度假区规划

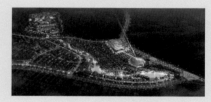

02 PUBLIC LANDSCAPE OF WANDA XISHUANGBANNA INTERNATIONAL RESORT 026
万达西双版纳国际度假区公共景观

03 HOTEL COMPLEX OF WANDA XISHUANGBANNA INTERNATIONAL RESORT 032
万达西双版纳国际度假区酒店群

04 BAR STREET OF WANDA XISHUANGBANNA INTERNATIONAL RESORT 054
万达西双版纳国际度假区酒吧街

05 XISHUANGBANNA WANDA PLAZA 058
西双版纳万达广场

C WANDA PLAZAS
万达广场
064

01 TAIYUAN LONGFOR WANDA PLAZA 066
太原龙湖万达广场

02 CHONGQING BA'NAN WANDA PLAZA 082
重庆巴南万达广场

03 DALIAN KAIFAQU WANDA PLAZA 090
大连开发区万达广场

04 SHANGHAI JINSHAN WANDA PLAZA 098
上海金山万达广场

05 GUANGZHOU LUOGANG WANDA PLAZA 104
广州萝岗万达广场

06 DONGGUAN HOUJIE WANDA PLAZA 110
东莞厚街万达广场

07 LIUZHOU CHENGZHONG WANDA PLAZA 118
柳州城中万达广场

08 GUILIN GAOXIN WANDA PLAZA 124
桂林高新万达广场

09 GUANGZHOU NANSHA WANDA PLAZA 130
广州南沙万达广场

10 NANNING ANJI WANDA PLAZA 136
南宁安吉万达广场

11 SICHUAN GUANGYUAN WANDA PLAZA 142
四川广元万达广场

12 NANTONG GANGZHA WANDA PLAZA 148
南通港闸万达广场

13 TAI'AN WANDA PLAZA 154
泰安万达广场

14 DEZHOU WANDA PLAZA 160
德州万达广场

15 DONGYING WANDA PLAZA 166
东营万达广场

16 HUANGSHI WANDA PLAZA 172
黄石万达广场

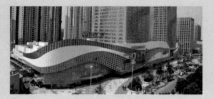

17 ZHEJIANG JIAXING WANDA PLAZA 178
浙江嘉兴万达广场

18 SUZHOU WUZHONG WANDA PLAZA 184
苏州吴中万达广场

19 FUYANG YINGZHOU WANDA PLAZA 190
阜阳颍州万达广场

20 NEIJIANG WANDA PLAZA 196
内江万达广场

21 QIQIHAR WANDA PLAZA 200
齐齐哈尔万达广场

22 ANYANG WANDA PLAZA 208
安阳万达广场

23 WEINAN WANDA PLAZA 214
渭南万达广场

24 YINGKOU WANDA PLAZA 218
营口万达广场

25 JIAMUSI WANDA PLAZA 222
佳木斯万达广场

D WANDA HOTELS 226
万达酒店

01 WANDA REIGN CHENGDU 228
成都万达瑞华酒店

02 WANDA VISTA HOHHOT 238
呼和浩特万达文华酒店

03 WANDA REALM LIUZHOU 246
柳州万达嘉华酒店

04 WANDA REALM TAI'AN 252
泰安万达嘉华酒店

05 WANDA REALM HUANGSHI 258
黄石万达嘉华酒店

06 WANDA REALM ANYANG 262
安阳万达嘉华酒店

07 WANDA REALM GUANGYUAN 266
广元万达嘉华酒店

08 WANDA REALM NEIJIANG 270
内江万达嘉华酒店

09 WANDA REALM DONGYING 274
东营万达嘉华酒店

10 WANDA REALM FUYANG 278
阜阳万达嘉华酒店

E DESIGN AND CONTROL 282
设计与管控

R&D AND APPLICATION OF "WANDA PLAZA STANDARDIZED DESIGN" 284
"万达广场标准化设计"的研发及应用

R&D AND APPLICATION OF "HUIYUN SYSTEM" V2.0 288
"慧云系统"（2.0 版）的研发及应用

F PROJECT INDEX 292
项目索引

A 序言
FOREWORD
WANDA COMMERCIAL PLANNING 2015

WANDA COMMERCIAL PLANNING 2015
万达商业规划 2015——持有类物业

SUCCESS OF CORPORATE TRANSFORMATION DAWNING
企业转型初见成效

服务业收入大幅增长

2015年万达服务业收入占集团收入达到43%,"服务业收入占比"比2014年提高10个百分点。房地产收入保持平稳,几乎没有增长。

商业租金大幅增长

商业租金增幅超过30%,"利润占比"预计超过35%。照此速度,很快万达商业"租金利润占比"就会超过50%。

轻资产打通道路

(1) 开启合作万达广场模式。以前万达都是自己买地自己投资,物业和租金全归自己。去年推出轻资产战略后,大批投资者上门,愿意出地出全部投资,万达出品牌,负责设计、建设、招商、运营;租金"七三分成",投资者占七成,万达占三成。这种分成比例在世界商业领域很罕见,一般做品牌管理,分成最多15%~20%。万达商业2015年已在北京、大连等地签约3个采用这种合作模式的项目。这让集团和万达商业管理层认识到,万达品牌是值钱的,决定将合作万达广场模式作为今后万达商业轻资产的主攻方向。这种模式有两大好处:一是零风险——地是净地,钱别人出,而且我们要求合作项目建设时,施工队伍必须是万达指定的队伍,所有设备必须使用万达品牌库中的优秀品牌,避免建设品质不好影响后期运营管理;二是不用资本化——只管建好管好项目,不用考虑资本化。

(2) 将建好的万达广场成本价卖给投资者,收回投资,租金也是投资者和万达"七三分成"。今年开业的25个轻资产万达广场,20个已签署协议。投资者为什么愿意接受"七三分成"?因为轻资产万达广场是中国市场上的稀缺资源,收入稳定、风险极低。轻资产万达广场的租金收益超过两位数,投资者表面看只拿到7%左右的回报率,但投资者自己买地投资可能还不如万达7%的租金回报率高。万达之所以能做到这么高的回报率,一是因为万达品牌有影响力,到哪里都受欢迎,拿地便宜;二是万达成批施工、集中采购,使建设成本大幅降低;三是重新设计了轻资产万达广场平面布局,以前万达广场的使用率55%左右,现在超过60%,增加了可租赁面积。

——摘自王健林董事长《万达集团2015年年度工作报告》

万达集团董事长
王健林

THE REVENUE FROM SERVICES ENJOYED SHARP INCREASE

In 2015, the yields from services contributed to 43% of Wanda Group's total revenue, up 10 percentage points year-on-year. The real estate revenue remained steady with flat increase.

THE RENTAL COLLECTED FROM COMMERCIAL PROPERTIES ROSE REMARKABLY

The rental from commercial properties grew over 30% with the profitability estimated to exceed 35%. Given such a speed, the profit from Wanda Commercial Properties will soon contribute more than 50% of the Group's total profit.

ASSET-LIGHT STRATEGY HAS FOUND ITS WAY

Firstly, partnership mode for Wanda Plazas kicked off. Wanda used to purchase land, invest and profit by its own. After Wanda announced its 'asset-light' strategy last year, many investors showed interest in providing land and fund, with Wanda providing the brand and taking charge of design, construction, tenant introduction and management, and both parties will share the rental at a 7:3 ratio: the investor takes 70%, and Wanda takes 30%, which is much higher than common practice with brand operators usually taking 15%-20% at most. In 2015, Wanda Commercial Properties signed three such projects repectively in Beijing, Dalian and another city, Which convinced the senior executives of the Group and Wanda Commercial Properties that the brand of Wanda is valuable, and it should be the major approach of asset-light strategy. The advantage of this approach are two folds: first, it will pose zero risk, since lands are clean and money from others, while Wanda oversees constructors and equipment during construction to avoid negative impact on operations due mal-quality; second, free from restraints of capitalization, which saves our efforts to concentrate more on development and operations.

Secondly, we sell Wanda Plazas after completion at the cost price to investors and recover the investment, and investors and Wanda will still share the rent at the ratio of 7:3. So far, we have signed 20 out 25 Wanda Plazas with the asset-light model that had opened in 2015. Why are investors willing to accept this ratio? The reason is that the Wanda Plazas are scarce assets in the Chinese market with stable revenue generation and very low risk. The asset-light Wanda Plazas can generate a two-digit rental profit margin. apparantly, an investor can get a return of about 7%, but if the investor buys land and invests by himself, he will not necessarily enjoy a higher rental profit margin than 7% offered by Wanda. Why can Wanda achieve such a high rate of return? Reasons are three folds: first, the brand is highly influential and widely accepted, which provides an edge to bargain for a lower price; secondly, with building in quantity and centralized procurement, costs drop sharply; thirdly, Wanda has redesigned the plan layouts of asset-light Wanda Plazas, and increased the utilization rate from 55% in the past to over 60% now, thereby increasing the floor space available for renting.

—— Excerpted from Chairman Wang Jianlin's Report on *2015 Wanda Group Work Report*

Chairman of Wanda Group
Wang Jianlin

WANDA COMMERCIAL PLANNING 2015
万达商业规划 2015

万达商业地产副总裁　赖建燕

2015年是万达集团正式对外宣布"第四次转型"的第一年。2015年万达商业地产全年开业项目40个，其中：万达广场26个，城市酒店10个，度假酒店3个，其他项目1个（万达云基地）；销售物业入伙75个，开放销售卖场137个。万达累计持有物业面积达2632万平方米，开业万达广场135个，酒店135家，总客房数21961间，继续保持全球领先地位。

万达商业规划系统在"新格局、新模式、新思维、新成果"的战略指导下开拓发展，在2015年创新性地推出、完成了多项具有企业里程碑意义的工作，对行业同样具有影响力。

一、提出并推动"项目BIM总发包管理模式"

2015年，全球BIM技术应用进入快速发展期，互联网技术日益成熟，万达项目管理工程"总包交钥匙模式"已经研发推行累积两年经验，万达国际项目投资管理也历经两年。此时，万事俱备并呼唤能够有一种新的"项目管理模式"来应对万达转型带来的更大规模化的投资建设！

2015年5月，万达集团丁本锡总裁采纳并批准了"商业规划"与"成本控制"两系统关于"将BIM技术全面引入项目投资管理的建议"，确定了万达"项目BIM总发包模式"的方向与内涵。万达"项目BIM总发包模式"是全球首个将BIM技术贯通于项目设计、建造、运维的全生命周期的项目开发智能化管理模式。这一模式被集团王健林董事长誉为"对全球不动产行业的革命性变革！"

万达商业规划系统首次代表集团牵头项目投资管理研发，与集团信息中心、成本控制部等8个部门组成联合研发小组，全面开展为期一年半的"BIM总发包管理模式"研发工作。其中"BIM总发包项目信息集成平台"等一系列科研成果将申报万达独立知识产权。万达"项目BIM总发包模式"计划将于2017年1月1日试运行（图1）。

万达"项目BIM总发包模式"的产品基础和技术基础，均是"万达广场"这个万达核心产品的标准化。

I. PUTTING FORWARD AND PROPELING "PROJECT BIM TURNKEY CONTRACT MANAGEMENT MODEL"

Year 2015 is the first year Wanda Group officially announced its "Fourth Transformation". In 2015, Wanda Commercial Estate opened 40 projects, among which there are 26 Wanda Plazas, 10 city hotels, 3 resort hotels and 1 other project (Wanda Cloud Base). In the same year, Wanda's properties for sale started 75 partnership projects and opened 137 sales stores. With the total property areas of Wanda adding up to 26.32 million square meters and 135 operating Wanda Plazas, 135 hotels and 21,961 guest rooms in total, Wanda has continued to be the global leader in the industry.

Under the strategic guidance of "New Vision, New Model, New Thinking and New Achievement", Wanda Commercial Planning System has enjoyed pioneering development in 2015, and has put forward and accomplished ingenious works of milestone significance inside the Group as well as influence to the industry.

In 2015, the application of BIM technology has entered into a repaid development period globally and the Internet technology has become increasingly improved. With two years of experience in the R&D work of "Turnkey Contract Model" by Wanda project management and two years of investment and management experience of Wanda international project department, Wanda is fully prepared at this moment and in anticipation of a new Project Management Model to cater for the investment and construction of greater scale as a result of Wanda's transformation plan.

In May 2015, Ding Benxi, CEO of Wanda Group, accepted and authorized the *Advice on Comprehensively Applying BIM Technology into Project Investment and Management* for the systems of Commercial Planning and Cost Control, which confirms the direction and connotation of Wanda's "Project BIM Turnkey Contract Model". Wanda's "BIM Turkey Contract Model" is the world's first intelligent management model of project development with BIM technology running through the full life cycle of projects (i.e. design, construction and operation & maintenance). This model is highly praised by Chairman Wang Jianlin as "A Revolutionary Transformation for the World's Real Estate Industry".

Headed by the Wanda Commercial Planning Sector on behalf of Wanda Group with input from eight departments (e.g. Wanda Information Center and Cost Control Department), a joint R&D team was established to carry out the R&D work on "BIM Turnkey Contract Model" in full swing for a year and a half, and Wanda will declare independent intellectual property rights for a series of research achievements, including the "Information Integration Platform of BIM Turnkey Contract Project". The model is planned to start pilot run on January 1, 2017 (Fig.1).

The foundation for the products and technologies of Wanda's "Project BIM Turnkey Contract Model" lies in the standardization of Wanda's core project - Wanda Plaza.

（图1）BIM总发包项目信息集成平台

《万达建造标准（2008版）》可以说是万达规划管理标准化的标志性起点，详细内容在本册《"万达广场标准化设计"的研发与应用》一文有展开性叙述。2015年万达商业规划在此标准化基础上开始的"万达广场标准版"的研究，是商业地产转型应对规模化发展的重要课题，也是"项目BIM总发包模式"科研的核心基础。

二、筹备"设计总包"，促进国内设计行业与国际接轨

2015年，在万达"总包交钥匙"项目管理的基础上，万达商业规划借鉴万达国际项目管理经验，完成了设计总包的全部筹备工作。万达"设计总包"管理制，将原万达广场"7+1"的规划设计发包管理归纳为"设计总包"一家综合管理，即极大地节省了万达规划管理成本，提高了设计和项目的管理效率；其大规模的实施，又促进了国内设计行业管理与国际的接轨，是国内设计行业管理进一步国际化、信息化的重大变革性举措。

三、完成全球首个企业"绿建节能五年规划纲要"

2011年，万达集团正式签发了第一个五年"节能规划纲要"——《万达集团节能工作规划纲要（2011—2015年）》。五年来，万达集团兑现对社会做出的承诺，提前、超额完成了"纲要"中能耗降低2%～3%的要求，年平均节能率大于5%。

截至2015年底，万达集团共计获得绿建认证标识359项，包括万达广场购物中心166项、酒店100项、住宅87项、文旅项目3项、万达学院2项和电商云基地1项，超额完成纲要目标（其中2015年新增"绿建认证标识"100项）。根据行业统计数据（《中国绿色建筑2015》，中国城市科学研究会主编），万达集团获得"绿建认证标识"的总数排名全国第一，超过第2名和第3名"认证"数量的总和。

随着万达业务在海外的开展，集团的绿建工作已涵盖了全球主流的"三大绿建认证体系"，包括美国LEED、英国BREEAM和澳大利亚GREENSTAR等。

2015年12月1日，万达集团总裁正式签批了由商业规划院"绿建节能研究所"牵头编制的第二个五年规划纲要——《万达集团绿建节能工作规划纲要（2016—2020年）》。新版"纲要"对集团整体绿建工作提出了更新、更高的要求，突出强调了高科技、智能化技术在绿建节能工作中的重要作用。

Wanda Construction Cost Standard (2008) marks the start of the standardization of Wanda's planning and management and an extensive narration of its detailed contents is included in the attached article - *R&D and Application of "Wanda Plaza Standardized Design"*. In 2015, on the basis of the above-mentioned standardization, the Wanda Commercial Planning started the research on the "Standard Edition of Wanda Plaza", which is a vital subject concerning the transformation of commercial property to cater for scale-up development as well as the core foundation for the research on the "Project BIM Turnkey Contract Model".

II. MAKING PREPARATION FOR "DESIGN TURNKEY" TO PROMOTE DOMESTIC DESIGN INDUSTRY AND MEET INTERNATIONAL STANDARDS

In 2015, on the basis of Wanda's "Turnkey Contract Project Management Model", Wanda Commercial Planning System drew lessons from the management experience of Wanda's international projects and finished all preparation work for "Design Turnkey". In Wanda's "Design Turnkey Management System", the original "7+1" planning and design contract awarding management model applied to Wanda Plaza was reduced to an integrated management by one company under the "Design Turnkey System", which greatly saves the cost of Wanda's planning and management and improves the efficiency of the management of design and project. In addition, its implementation on a large scale propels the management of the domestic design industry to act on international standards. "Design Turkey" is seen as an important revolutionary measure taken by the management of the domestic design industry to further promote its internationalization and informationization.

III. COMPLETING THE FIRST "GREEN BUILDING ENERGY-SAVING FIVE-YEAR WORK PLAN"

In 2011, Wanda Group officially issued its First Energy-saving Five-Year Planning Outline , i.e. *Wanda Green Building Energy-saving Work Plan (2011-2015)*. In the past five years, Wanda Group has honored its commitment to the society and out performed the requirement in the Plan of cutting down energy consumption by 2%-3% in advance, with an annual average rate of energy saving larger than 5%.

By the end of 2015, Wanda Group has obtained 359 Green Building Labels in total, including 166 for Wanda Plaza shopping centers, 100 for hotels, 87 for residential buildings, 3 for cultural tourism projects, 2 for Wanda Institutes and 1 for Wanda E-commerce Cloud Base and outperformed the objective noted in the Plan (the newly increased Green Building Labels in 2015 amount to 100). According to the industrial statistic data - *China Green Building 2015* compiled by Chinese Society for Urban Studies, Wanda Group topped China national Green Building Label list, exceeding the aggregate label number of the companies that rank number two and number three.

Along with the expansion of Wanda's overseas business, Wanda's Green Building Label has covered the three global mainstream green building label systems, including LEED (USA), BREEAM (UK) and GREENSTAR (Australia).

On December 1, 2015, Chairman Wang Jianlin officially approved the second Five-Year Plan - *Wanda Green Building Energy-saving Work Plan (2016-2020)* led by the Green Building Energy-saving Research Department at Wanda Commercial and Planning Institute. The new plan has posed newer and higher requirements on Wanda's Green Building work, and meanwhile, stressed the significant role of high technology

万达集团作为一个民营企业,一直将绿建节能、可持续发展作为集团的顶层战略目标,并通过制定"五年节能规划纲要",推进绿建节能工作的开展。2001年的"大连雍景台项目"外墙外保温新工艺走在了行业的前列,2002年的"昆明滇池卫城项目"是云南省首个进行环境评估的住宅项目。2010年开始,万达集团的绿建节能工作进入体系化战略发展阶段,"五年节能规划纲要"的制定和实施,充分体现了万达集团的企业社会责任!

四、牵头研发升级"慧云"(2.0版)并实施验收

2015年,万达商业地产首次成立"慧云专项工作小组",由万达商业规划研究院、万达商业管理有限公司、万达集团信息管理中心、万达商业地产成本部等多部门专职人员组成。2015年"慧云专项工作小组"完成了"慧云"(2.0版)的全部研发工作并付诸实施。截至2015年底,已有62个项目建成"慧云"(2.0版)系统,在不降低运行安全性、不改变操作便利性的前提下,优化了系统,大大降低了现场实施难度,提高了验收的考核标准,能够更好地满足"商管"的使用要求。

"慧云"1.0版和2.0版的实施,可为每个万达广场减少管理人员12名,每年直接降低运行管理成本百万元,为大型商业广场的绿色运营、智能化运营做出了表率(图2)。

万达商业规划在2015年完成了商业规划"设计总包"的全部准备工作。万达与各设计合作伙伴的关系更加密切,管理界面更加清晰,工作效率更高,产品成果也更加成熟和富有品质。

2015年开业的"西双版纳国际度假区"是万达第一个带"主题公园"和"秀场"的文化旅游项目,与长白山国际度假区、武汉中央文化区被王健林董事长称作万达文化旅游类"第一代产品"。其中由万达商业规划院总体规划,与万达酒店设计院等设计公司联合设计的西双版纳文华酒店开业后荣获多项国际大奖(图3)。2015年开业的"持有物业"中成都瑞华酒店是万达第二座等级酒店品牌"瑞华"酒店,太原万达广场是2015年开业的唯一一个"A级店"商业广场。"销售物业"中,成都万达城展示中心采用"芙蓉花"主题造型,虽然文化旅游项目尚在建设中,但已吸引了众多的游者和顾客。成都万达城"120合院别墅"(图4)、"版纳90墅"等一批创新类产品,也赢得了市场的好评。

2015年万达商业规划共获得4项国际大奖,37项国

and intelligent technology in Green Building Energy-saving work.

As a private enterprise, Wanda Group has always taken green building, energy saving and sustainable development as its top strategic objectives of the Group and promoted the development of green building and energy saving work by formulating the Five Year Energy-saving Work Plan. The external wall thermal insulation technology applied in the Dalian Robinson Place Project in 2001 is a pioneering new technology in the industry; the Kunming Dianchi Lake Project of 2002 is the first residential project with environment assessment in Yunnan Province. The green building and energy saving effort of Wanda Group has entered into a systematic and strategic development phase from 2010, and the formulation and implementation of the Five-Year Energy-saving Work Plan fulfill the enterprise's commitment to shouldering social responsibility.

IV. LEADING THE RESEARCH AND DEVELOPMENT OF UPGRADING "HUIYUN SYSTEM"(V2.0) AND ACCEPTANCE

In 2015, a "Dedicated Huiyun System Work Team", consisted of full-time staff from various departments including Wanda Commercial Planning Research Institute, Wanda Commercial Management Limited Company, Wanda Group Information Management Center and the Cost Control Department of Wanda Commercial Properties was established by Wanda Commercial Properties for the first time. In 2015, the complete R&D work of "Huiyun System"(V2.0) was completed and put into practice by the team. As of the end of 2015, 62 projects have configured "Huiyun System"(V2.0). Without compromising on operation safety and operation convenience, "Huiyun System"(V2.0) has optimized the system, significantly reduced onsite implementation difficulty and improved assessment criteria for acceptance, better catering for operation requirements of the Commercial Management Department.

With the implementation of "Huiyun System"(V1.0) and "Huiyun System"(V2.0), each Wanda Plaza can cut down 12 management staff, which will directly reduce the operation and management cost by RMB one million a year. It sets an example for the green and intelligent operation of large commercial plazas (Fig.2).

In 2015, Wanda Commercial Planning has finished all preparation work for the "Design Turnkey" of its commercial planning. Wanda and its design partners have become increasingly closer, with clearer management interface,

(图2)"慧云2.0"智能化管理机房

(图3)西双版纳万达文华酒店观景平台

（图4）成都万达城"120合院别墅"

内大奖（其中全国人居经典建筑规划设计方案竞赛奖项26项，国内其他奖项11项），同时获得国家专利21项。

2015年，万达商业规划作为万达商业地产核心产品的研发部门，在万达的"第四次转型"中，担负着所有产品转型创新的重任，必须在有限的时间内，应对市场剧烈变化所带来的各类风险和挑战。万达商业规划在面对市场的多重不确定性及整体宏观把控方面，得到了万达集团董事长、总裁的悉心指导，得到了各相关部门的鼎力协助与支持。可以说，没有万达的企业文化，就不可能在极短的时间内率先完成"万达转型"中的"规划设计转型"；也就不可能完成规划设计来年开业55个万达广场、两个万达城的艰巨任务；更不可能实现下半年以BIM技术为基础的项目"BIM总发包管理模式"的牵头研发工作。

2015年是万达商业规划走向成熟稳健的一年。本"年鉴"首次将万达年度开业的"持有类物业"和"销售类物业"合集出版，其中"持有类物业"包括2015年全部开业项目，"销售类物业"包括部分入伙项目。2015年之后，由于万达的"第四次转型"，万达商业地产的产品也将增加"投资主体类别"的划分维度，万达广场将分为包括销售物业的"开发项目"，自行投资不含销售物业的"直投项目"以及与其他公司合作的"合作项目"三大类型；万达城也将出现"开发项目"与"合作项目"两大类型。2015年，在万达集团的战略转型过程中，必将是里程碑式的一年，是万达的转型年。

elevated work efficiency and more mature and quality product outcome.

The Xishuangbanna International Resort opened in 2015 is Wanda's first cultural tourism project with theme park and Show Theater. It is categorized as the "first generation product" of Wanda cultural tourism project by Chairman Wang Jianlin along with the Changbai Mountain International Resort and Wuhan Central Culture District. The Wanda Vista Resort Xishuangbanna, planned by Wanda Commercial Planning Institute and jointly designed by Wanda Hotel Design Institute and other design companies, has won a number of international awards after its operation (Fig.3). Among Wanda's properties for holding opened in 2015, Wanda Reign Chengdu is the second Reign Hotel of Wanda and Taiyuan Wanda Plaza is the only A-level flagship commercial plaza that opened business in 2015. Among Wanda's properties for sale, the Lotus design scheme is adopted by the Chengdu Wanda Cultural Tourism City Exhibition Center. Although other cultural tourism projects are still under construction, many tourists and customers have already been attracted. A series of innovative projects, including the 120-Courtyard Villa of Chengdu Wanda City (Fig.4) and 90-Villa of Xishuangbanna, have also won favorable reception from the market.

In 2015, Wanda Commercial Planning has won four international awards, 37 national awards (including 26 awards in National Human Settlements Architecture & Planning Competition Awards and 11 other awards) and 21 national patents.

In 2015, as a department engaged in R&D of the core products of Wanda Commercial Properties, Wanda Commercial Planning has shouldered the important responsibility of the transformation and innovation of all products in the context of the Fourth Transformation of Wanda Group. It has to cope with various risks and challenges brought by drastic market changes within limited time. In the aspects of confronting the multiple uncertainties of the market and the overall macroscopic control, Wanda Commercial Planning System has received meticulous guidance from the Chairman and the CEO of Wanda Group as well as the great assistance and coordination from relevant departments. It is fair to say that without the enterprise culture of Wanda, it's impossible to finish the task of planning and design transformation set forth in the transformation of Wanda in such a short time; it is also impossible to accomplish the daunting task of opening 55 Wanda Plazas and 2 Wanda Cities in the following year as planed; still less to realize the R&D work of the "BIM Turnkey Management Model" based on BIM technology in the second half year.

The year 2015 is a year that Wanda Commercial Planning System moves towards maturity and steadiness. In this year book, the "Properties for Holding" and "Properties for Sale" of Wanda opened in this year are published collectively for the first time, in which the properties for holding include all projects opened in 2015 and properties for sale incorporate partial partnership projects. After the year 2015, given the fourth transformation of Wanda, the dividing dimension of "Investment Subject Category" will be added to the products of Wanda Commercial Properties. Wanda plaza projects will be categorized into three types, which are the development projects of properties for sale, the direct investment projects without properties for sale and the partnership projects in cooperation with other companies; Wanda city projects will also have two types, which are development projects and partnership projects. In the strategic transformation process of Wanda Group, the year 2015 will certainly be a year of milestone significance as well as a year of transformation.

B
WANDA XISHUANGBANNA INTERNATIONAL RESORT
万达西双版纳国际度假区

01 WANDA XISHUANGBANNA INTERNATIONAL RESORT PLANNING
万达西双版纳国际度假区规划

规划时间：2010 / 03 – 2015 / 09
规划位置：云南 / 景洪
规划面积：524 公顷
建筑面积：343 万平方米
规划功能：主题乐园、秀场、万达广场、酒店群、旅游新城、"三甲"医院等

PLANNING PERIOD : MARCH 2010 TO SEPTEMBER 2015
LOCATION : JINGHONG, YUNNAN PROVINCE
PLANNED AREA : 524 HECTARES
FLOOR AREA : 3,430,000M^2
FUNCTIONS : THEME PARK, SHOW THEATER, WANDA PLAZA, HOTEL COMPLEX, NEW TOURISM CITY, GRADE-A CLASS-3 HOSPITAL, ETC.

一、引言：万达西双版纳国际度假区——世界级旅游度假项目

度假休闲游是一种新兴的旅游模式。相比观光旅游等传统旅游产品，度假休闲旅游目的地的主要特点是：具有"优越生态环境"、"优美景观环境"、"优雅文化环境"和"优良服务环境"的"四优"度假区，提供游客全方位的到达、停留、休闲和体验服务。

2015年我国人均GDP已突破8000美元，度假旅游需求持续高速增长。万达西双版纳国际度假区正是应运这种市场趋势，由万达集团倾情打造的高端度假休闲目的地。它不仅是西南地区投资额最大文化旅游项目，同时也是内容最丰富、业态最创新的世界级旅游度假项目。

二、万达城——万达集团商业地产转型、创新之集大成者

万达集团于2006年战略重心向文化、旅游产业转移。针对文化旅游业，万达的核心产品是万达文化旅游城。万达文化旅游城是万达集团凭借多年在商业、文化、旅游产业积累的丰富经验，整合全球资源，创新应用世界一流的科技设备和手段，倾力打造的创意独特并突出中国元素的世界级特大型文化旅游项目。

截至目前，万达城经历了从第一代至第四代的发展历程。其中第一代为传统的休闲度假区，以长白山为代表，依托滑雪项目，建设有度假酒店、旅游小镇等，总体以常规休闲度假为主；第二代为综合文化旅游区，具备部分城市级配套功能，以西双版纳国际度假区、武汉中央文化区为代表，包含主题乐园、秀场、酒吧街等部分主题娱乐业态及商业街、商务区等部分城市配套功能；第三代为全业态、具备完整城市级配套功能的万达城，涵盖万达茂、主题乐园、秀场等多种主题娱乐业态和医院、学校等城市配套功能，以南昌万达城、合肥万达城、青岛东方影都为代表；第四代为具备产业功能的万达城，以印度产业新城项目为代表。

万达城作为万达独创的文化旅游项目，与其他文化旅游项目（如国内的华侨城、长隆及国际著名的迪士尼、环球影城等）相比，具备以下特点：
（1）多功能复合——大部分的文化旅游项目业态单一，主要业态为主题乐园、配套酒店。万达城则除了包含中国元素与地域文化完美融合并引入高科技的主题乐园，还包括万达独创的以室内游乐项目与商业结合的核心产品万达茂，顶级度假酒店、秀场、酒吧街等，致力于打造全业态、功能复合的超大型综合高端文化旅游度假目的地。
（2）城市级配套——万达城还提供完善的城市级功能服务，包括生态社区、商业街区、高端医院、一站式学校等，与城市发展需求高度契合。
（3）地域性文化——大部分主题乐园到任何地方都是一个IP为主，强调其可复制性，而每个万达城都深度挖掘当地地域文化，创造一个具有浓郁中国元素和地域文化特色的独一无二的文化旅游度假区。

西双版纳国际度假区作为万达集团开业的第二代文化旅游项目，集团倾力打造，成为文化旅游度假区的典范。实现了全球首创的"全业态、大规模、集群化"发展模式，构建了充满活力、功能复合、地域文化彰显、创新发展的旅游度假区。

I. INTRODUCTION : WANDA XISHUANGBANNA INTERNATIONAL RESORT - A WORLD-CLASS TOURISM AND RESORT PROJECT

Recreation tourism is a newly emerging travel model. In comparison with the traditional tourism product like sightseeing travelling, the destinations of recreation tourism are featured with "Four Excellence" resorts which are "Excellent Ecological Environment", "Excellent Landscape Environment", "Excellent Cultural Environment" and "Excellent Service Environment". They are offering tourists with full aspects services in the processes of arriving, staying, travelling and experiencing.

In 2015, China's per capita GDP surpassed 8,000 USD and the demand of recreation tourism witnesses a sustainable and speedy growth. In respond to such a market trend, Wanda Group puts forth efforts in building the Wanda Xishuangbanna International Resort, a high-end destination for vacation and recreation. Being the largest cultural tourism project in Southwest China in terms of investment value, it is also a world-class tourism and resort project with the most diversified programme and innovation business types.

II. WANDA CITY - AN EPITOME OF TRANSFORAMTION AND INNOVATION OF WANDA COMMERCIAL PROPERTIES

From 2006, Wanda Group has strategically transferred its focus to the culture and tourism industry. Targeting at the cultural tourism industry, Wanda launched its core product - Wanda Cultural Tourism City. Depending on the rich experiences accumulated in the commercial, cultural and tourism industry, integrating global resources and using state-of-the-art innovative science and technology equipment and methods, Wanda Group has strived to build the Wanda Cultural Tourism City, a world-class super large cultural tourism project with unique originality and distinctive Chinese elements.

Up to present, Wanda City has gone through a development history from the first generation to the fourth generation. The first generation refers to the traditional recreation resorts represented by the Changbai Mountain International Resort which relies on skiing, resort hotel and tourism town with the main purpose of regular leisure and recreation. The second generation refers to the comprehensive cultural tourism areas and districts which are equipped with partial city-level supporting functions. The typical representatives include Xishuangbanna International Resort and Wuhan Central Cultural District, which include theme entertainment business types such as theme park, Show Theater, bar street and some city supporting functions like Commercial Street and business district. The third generation refers to the Wanda City with integral business types and complete city-level supporting functions, covering various theme entertainment business types like Wanda Mall, theme park and Show Theater and city supporting functions like hospital and school. The typical representatives include Nanchang Wanda City, Hefei Wanda City, and Qingdao Oriental Cinema. The fourth generation refers to Wanda City with industrial functions which are represented by the India Industrial New City.

As a cultural tourism project originally created by Wanda, comparing with other cultural tourism projects such as the Overseas Chinese Town, Changlong Park, the internationally renowned Disney Land and Universal Studio, Wanda City has the following characteristics:
(1) Combination of multiple functions: most cultural tourism projects have few programme with their major business types of theme park and supporting hotel. While, Wanda City includes not only theme park with a perfect integration of Chinese elements and local culture as well as the application of high-tech, but also the Wanda Mall, a core product originally created by Wanda that combines indoor theme park, exclusive resort hotel, Show Theater, bar street, etc., striving to create an ultra-large comprehensive high-end cultural tourism destination with full aspect programme and functions.
(2) City-level supporting programme: Wanda City also provides sophisticated city-level functional services, including ecological community, commercial district, high-end hospital, one-stop school, etc., which meets seamlessly with urban development demand.

三、西双版纳国际度假区——国际一流文化旅游度假区

万达西双版纳国际度假区位于云南省西双版纳傣族自治州景洪市西北部,占地5.3平方公里,地理位置优越,处于昆(明)曼(谷)国际大通道旅游线和有"东方多瑙河"之称的澜沧江—湄公河流域交叉点。其交通便捷,距景洪市中心3千米,距西双版纳景洪国际机场约4.6千米,距景洪港(澜沧河—湄公河航道上我国境内第一大港口)约6千米(图1)。

万达西双版纳国际度假区由万达商业规划研究院整体规划设计,历时3年,数易其稿,于2012年正式开工,2015年9月全面建成开业。万达西双版纳国际度假区应用高科技设备和手段,围绕"热带雨林"和"傣族风情"两个核心元素进行设计,精心打造了呈现原生态的雨林生活和傣族神话的"傣秀"、展现傣族建筑特色和南传佛教文化的"山地度假酒店群"、"文化主题乐园"等独具地域文化特色的文化旅游项目,构筑西南地区第一条完整的"文化旅游产业链"(图2)。

四、"交通优先、功能复合、突出文旅、地域文化"四位一体,完美打造度假区

1. 交通优先——区位优势强化,交通导向为主,打造道路山水骨架,串联各大功能区

(3)Vernacular culture: most theme parks, no matter where they are, are similar with one same IP which emphasizes the reproducibility. While each Wanda City, by probing deeply into the local and regional culture and creating cultural tourism resort with rich Chinese elements and local cultural characteristics, is ensured that it is of great importance and uniqueness.

Xishuangbanna International Resort, the second generation cultural tourism project meticulously developed by Wanda Group, has set an outstanding model for cultural tourism resorts. It introduces the internationally pioneering development model of "full formats, large scale, clustering development", and thus builds a tourist resort which boasts exuberant vitality, composite functions, rich regional culture and innovative development.

III. XISHUANGBANNAN INTERNATIONAL RESORT - A WORLD LEADING CULTURAL TOURISM RESORT

Located in the northwest of Jinghong, Xishuangbanna Dai Autonomous Prefecture covering a site area of 5.3 square kilometers, Wanda Xishuangbanna International Resort enjoys a good geographical location, nestled at the intersection of Kunming-Bangkok International Corridor Tourism Route and the Lancang-Mekong River Basin, which earns a reputation of the Oriental Danube. It also enjoys a convenient transportation: 3km away from downtown Jinghong City, 4.6km away from Xishuangbanna Jinghong International Airport and 6km away from Jinghong Port, which is the biggest port within Chinese board in the Lancang-Mekong River channel (Fig.1).

It took Wanda Commercial Planning Research Institute three years to finish the overall planning and design of Wanda

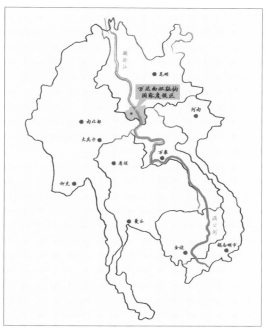

(图1)西双版纳国际度假区区位示意图

(图2)万达西双版纳国际度假区东区鸟瞰图

（1）便捷高效的区域交通——度假区通过西双版纳国际机场、泛亚高铁、景洪港可便捷到达全国各地，并辐射东南亚地区。
（2）舒适顺畅的内部交通——度假区道路依山就势，形成以A1路为主、环状路为辅的布局结构，将各个功能区联系起来。
（3）先进宜人的慢行系统——整体结合网状的绿地和道路系统、几大湿地公园和山地度假酒店区设置多重专用人行慢跑道，形成步移景异的多重健康慢行系统（图3）。

2. 功能复合——文旅功能为核，提供完善的城市级功能配套

（1）文旅功能为核——度假区汇聚了文化旅游度假的创新核心产品，包括室外主题乐园、傣秀剧院、高端度假酒店群、风情滨湖酒吧街、万达主题乐园、雨林体育公园等文化旅游项目，给人们提供多方位的娱乐体验。
（2）城市级功能配套——度假区还提供一流的城市级配套和完善的公共服务，包括宜居的生态社区、成熟的商业街区、高水平的"三甲"医院、"一站式"的中小学教育设施、大型生态湿地公园以及消防站、公安局等城市保障设施，共同构建成充满活力、功能复合创新的旅游度假区。
（3）精彩纷呈八大区，活力国际度假区——度假区建设有万达主题乐园、傣秀剧院、高端度假酒店群、

Xishuangbanna International Resort. In these three years, the design drawings were repeatedly revised. The construction work of the Resort was formally commenced in 2012, and it was opened for business in September 2015. With the application of high-tech equipment and methods the design of Wanda Xishuangbanna International Resort centers on the two core elements of "Tropical Rainforest" and the "Dai Ethnic Folklore". Meticulous efforts have been made to create cultural tourism projects with unique local and cultural characteristics, i.e. Dai Show presents the original rainforest life and the mythology of the Dai nationality, the Mountain Resort Hotel Complex brings the architectural features of the Dai nationality and the Southern Buddhism culture, and the cultural theme park, which contributes to the creation of the first integrated cultural tourism industrial chain (Fig. 2).

IV. CREATING A PERFECT RESORT WITH FOUR ELEMENTS IN ONE: "ACCESSIBLE TRAFFIC, COMPREHENSIVE FUNCTIONS, HIGHLIGHT CULTURAL TOURISM AND RICH REGIONAL CULTURE"

1. ACCESSIBLE TRAFFIC: INTENSIFY REGIONAL ADVANTAGES WITH PRIORITY GIVEN TO TRAFFIC GUIDE TO CREATE ROAD PHYSIOGNOMY FRAMEWORK AND CONNECT THE MAJOR FUNCTIONAL AREAS

(1) Convenient and efficient local transportation: the Resort can be easily accessed from all parts of the country, and even from the Southeast Asia Region via the Xishuangbanna International Airport, the Pan-Asia High-speed Rail and the Jinghong Port.

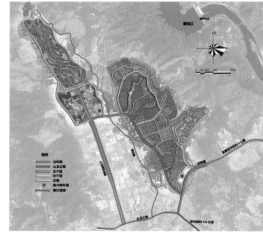

（图3）道路交通分析图

(2) Comfortable and smooth internal transportation: the roads inside the Resort are constructed along the hills, forming an A1 road-oriented and loop-oriented layout system that link the main functional areas.
(3) Advanced and agreeable slow-traffic system: multiple dedicated jogging pavements are arranged with an overall consideration of the netted greenbelt and road system, the wetland parks and the mountain resort hotel area to form a multi-layered jogging system that presents varying sceneries with changing view-points (Fig. 3).

2. COMPOSITE FUNCTIONS: PROVIDE PERFECT URBAN-LEVEL MATCHED FUNCTIONS WITH CULTURAL TOURISM FUNCTION AS THE CORE

(1) Core cultural tourism function: the Resort gathers the core and innovative products of cultural tourism, including outdoor theme park, the Dai Show Theater, high-end resort hotel complex, Stylish Lakeshore Bar Street, Wanda Theme Park and Rainforest Sports Park, and offers people multi-directional entertainment experiences.
(2) City-level supporting functions: the Resort also provides first-rate city-level sophisticated supporting public services, including livable ecological community, mature commercial block, high-level Grade-A hospital, one-stop primary and secondary school education facilities, large-scaled ecological wetland park and city security facilities such as fire station and public security bureau, which jointly create a vital tourist resort with composite and innovative functions.
(3) Dynamic international resort with eight brilliant areas: the Resort consists of eight function areas, which are Wanda Theme Park, Dai Show Theater, High-end Resort Hotel Complex, Bar Street, Commercial Center, Large-scaled Rainforest Sports Park, New Tourism City and Grade-A Hospital. Detailed descriptions are listed as follows (Fig. 4).

A. WANDA THEME PARK - RICH IN RENGIONAL CULTURE AND PROVIDES VITAL EXPERIENCES
Located in the west wing of the Resort, the Wanda Theme Park (covering a site area of 60 hectare) is designed by the FORREC, the first rated design firm of such kind in the world. As a development focus of cultural tourism, the Theme Park is rich in cultural features of the Dai Nationality, consisting of five scenic regions consisting of the Butterfly Kingdom, the Jungle Adventure, the Ancient Tea Horse Road, the Fisherman's Wharf and the Water Park. It can satisfy the demand of different tourists, with a designed yearly reception capacity of 2.5 million persons (Fig. 5).

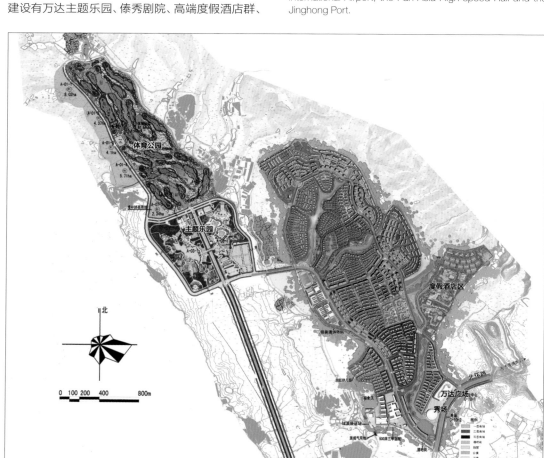

（图4）万达西双版纳国际度假区总平面图

(图5) 万达西双版纳国际度假区主题乐园实景鸟瞰图

(图6) 万达西双版纳国际度假区傣秀剧场

酒吧街、商业中心、大型雨林体育公园、旅游新城和"三甲"医院等八大功能区，具体如下（图4）。

A. 万达主题乐园——地域文化，活力体验

万达主题乐园位于度假区"西翼"，占地60公顷，由世界著名的FORREC公司担纲设计。主题乐园作为文化旅游的发展重点，极具傣族文化特色，由蝴蝶王国、丛林冒险、茶马古道、渔人码头和水乐园"五大景区"构成，可以满足不同游客的需求，设计年接待量250万人次（图5）。

B. 傣秀剧院——大师设计，艺术体验

傣秀剧院紧邻北环路，建筑面积2万平方米。建筑设计由国际大师马克·费舍尔先生担纲设计。秀场建筑外形平缓尖顶的结构源于棕榈叶的几何形状，将时尚与地方特色完美结合，是国际顶级现代科技剧院。傣秀由世界顶尖艺术创作大师弗兰克·德贡先生策划导演，将民族艺术，结合科技的灯光音效、舞蹈、杂技、跳水、沙画等多种表演形式，呈现出一台美轮美奂的舞台艺术，达到世界一流演出水准（图6）。

C. 高端度假酒店群——山地度假，文化体验

度假酒店群位于度假区门户位置，由万达商业规划院精心设计，占地约29公顷，依山而建，规划建设3个度假酒店，包括万达文华酒店（六星级）、万达希尔顿逸林度假酒店（五星）和万达皇冠假日度假酒店（五星），共有985间客房，约1500个床位，充分满足游客的需求。文华酒店西侧规划有西双版纳总佛寺禅修堂，依托酒店又相对独立，植物多采用与南传佛教有关的"五树六花"；东北侧是令人身心放松的SPA区，依山傍水，坐听溪流潺潺，修身静心（图7）。

D. 滨湖酒吧街——乐享生活，休闲体验

滨湖酒吧街滨水而建，建筑面积1.5万平方米，引进多家国内外著名酒吧、音乐吧品牌，乐享现代滨水休闲生活方式。

E. 万达广场——时尚创新，汇聚精品

万达广场位于度假区东侧门户位置，建筑设计融合了本地民族特色的孔雀元素，将城市商业繁华与地方文化充分融合。建筑面积7万平方米，引进超市、万达影城与两百多个国内外知名商家，集时尚、购物、休闲、娱乐、美食于一体，为本地居民和国内外游客提供国际一流的休闲购物体验，是西双版纳最具特色、最具现代感、品质感、最惬意的商业中心（图8）。

F. 大型雨林体育公园——体育竞技，雨林体验

大型雨林体育公园位于度假区"西翼"最端头，占地90公顷，由世界顶级的设计大师罗伯特·琼斯先生担纲设计。雨林体育探险公园紧邻自然保护区，植被丰厚，自

B. DAI SHOW THEATER - A WORK OF DESIGN MASTER THAT PROVIDES ARTISTIC EXPERIENCE

Located closely to Beihuan Road, the Dai Show Theater covers a gross floor area of 20,000 square meters. Its architectural designer, Mr. Fisher, is a highly reputed international master. The Show Theater adopts a gentle pinnacle structure in its architectural appearance, which is an inspiration derives from the geometrical shape of palm leaf. With a perfect combination of fashion and local features, it lives up to the reputation of an international top-level modern technology theater. The Dai Show is directed by the world-leading musical creative master Mr. Frank Dagon, who combines the Chinese national art with technological lighting and sound effect as well as various forms of performance like dancing, acrobatics, diving and sand painting showing magnificent stage art which reaches a world-class standard (Fig. 6).

C. HIGH-END RESORT HOTEL COMPLEX - VACATION IN MOUNTAIN WITH CULTURAL EXPERIENCE

Located in the gateway of the resort, the resort hotel complex is meticulously designed by Wanda Commercial Planning Institute. Covering a site area of 29 hectares and built along hills, three resort hotels: WANDA VISTA RESORT XISHUANGBANNA (six stars), DOUBLETREE RESORT BY HILTON HOTEL XISHUANGBANNA (five stars) and CROWNE PLAZA RESORT XISHUANGBANNA PARKVIEW (five stars) are planned and built, fully satisfying visitors' needs with 985 guestrooms and about 1,500 beds. The Meditation Hall Pagoda of Xishuangbanna Head Buddhist Temple is planned to be built to the west of the Vista Hotel, which relies on the hotel with relative independence. The plants are mainly arranged as per the belief of "Five Trees and Six Flowers" related to the Southern Buddhism; the SPA area lies in the northeast wing, which is near the mountain and by the rivers. Listening to the murmuring stream, visitors can cultivate their moral character and calm their minds (Fig. 7).

D. LAKESHORE BAR STREET - ENJOY LIFE WITH RELAXING EXPERIENCE

The Lakeshore Bar Street is built by the riverside, with a gross floor area of 15,000 square meters. Various domestically and internationally renowned brands of bars and music bars are brought in to give visitors the enjoyment of modern lakeshore relaxing life style.

E. WANDA PLAZA - FASHION AND INNOVATION CONVERGENCE BOUTIQUES

Located in the eastern gateway of the resort, the architectural design of Wanda Plaza fuses the peacock element of the local folk culture, which fully integrates the prosperity of urban commerce with the local culture. With a gross floor area of 70,000 square meters, supermarket, Wanda Cinema and more than 200 renowned domestic and international shops are brought in to provide a world-leading relaxing shopping experience, integrating fashion, shopping, relaxation, entertainment and catering, for local citizens and visitors home and abroad, making it the coziest business center with most distinctive features, strongest modern sense and highest quality (Fig. 8).

F. LARGE-SCALE RAINFOREST SPORTS PARK - SPORTS COMPETITION WITH RAINFOREST EXPERIENCE

Located in one end of the west wing of the Resort and covering a site area of 90 hectares, the large-scaled Rainforest Sports Park is designed by a world-leading designer, Mr. Robert Jones. Located closely to the Nature Reserve, the Rainforest Sports Adventure Park is rich in vegetation with beautiful natural environment, serving as the best example of rainforest vacation concept of Wanda International Resort. Combining sports competition with rainforest experience, it enables visitors to experience the passion of sports with a feeling of original rainforest, and is a window to show the features of the local landscape and the theme of fitness and health-keeping.

（图7）万达西双版纳国际度假区酒店群

（图8）西双版纳万达广场

然环境优美，是雨林度假理念在万达国际度假区的最好展现。其将体育竞技与雨林体验结合在一起，在原始雨林风情中感受体育动感，是展示地域景观特征和康体健身主题的窗口。

G.旅游新城——活力宜居，配套完善

旅游新城占地约240公顷，规划提供了多种类型的公寓、洋房、联排和别墅，以满足不同的度假居住需求，打造配套齐全、环境优美、充满活力的居住环境。旅游新城还规划建设了幼儿园、小学、中学等完整的教育配套设施，可以享受"一站式"的教育服务体系。

H."三甲"医院——完善医疗保障

万达西双版纳国际度假区规划建设了一座"三级甲等"医院（云南省第一人民医院西双版纳医院），提供国际标准的医疗保障，为景洪及周边的地区人民及游客提供多层次的诊疗服务，极大地提高城市医疗服务水平和就医环境。

3.突出文旅——打造"一核"、"两翼"、"三带"，营造山水绿色之城

项目从规划之初到最终定稿，历经3年，数易其稿，开发建设用地面积也从最初的6.1平方千米减少到5.3平方千米。其原因就在于充分尊重基地的山水、自然环境，尽量减少对环境和原有生态系统的影响，利用GIS等先进手段，不断优化方案，严格控制各种生态

G. NEW TOURISM CITY - DYNAMIC AND LIVABLE WITH COMPLETE SUPPORTING SERVICES

Covering a site area of 240 hectares, the planning of the New Tourism City involves various types of apartments, western-style houses, townhouses and villas to cater to different holiday living demands and create a dynamic living environment with complete supporting services and beautiful environment. A complete set of supporting education facilities including kindergarten, primary school and secondary school is also planned to be built in the New Tourism City, providing a "one-stop" education service system.

H. GRADE-A HOSPITAL - IMPROVE HEALTH CARE PROTECTION

A Grade-A hospital i.e. Xishuangbanna Hospital of Yunnan No.1 People's Hospital, is planned and built in the Wanda Xishuangbanna International Resort, providing health care of international standard and multi-layered treatment services for citizens in Jinghong and surrounding regions as well as visitors, which significantly improves the medical service level and hospital environment of the city.

3. HIGHLIGHT CULTURAL TOURISM: A CITY WITH GREEN LANDSCAPE IS CREATED VIA THE CONSTRUCTION OF ONE CORE AREA, TWO WINGS AND THREE BELTS

From the beginning of the project planning, it took three years to finalize the design drawings of the project, which were repeatedly revised over the three years. The development and construction site area is also reduced from the original 6.1 square kilometers to 5.3 square kilometers. The purpose is to fully respect the mountains, rivers and natural environment of the site and to minimize the impact to the environment and the original ecosystem. With the utilization of advanced technology like GIS, the scheme is continuously optimized to strictly control the site area of various ecological corridors and build excellent ecological environment in the Resort.

Under limited conditions, a landscape and environment with distinctive spatial features is created (Fig. 9).

Meanwhile, more importantly, the theme of cultural tourism is highlighted by the overall planning and arrangement to make sure that the circulation and layout of cultural tourism agree with visitors' need with the greatest conformity to objective laws. Taking road landscape as the skeleton, the main function areas spread in sequence along the A1 main road with business types of public function and cultural tourism arranged at the two ends of the site. To be more specific, a city with green landscape is created via the construction of "One Core", "Two Wings" and "Three Belts".

(1) "One Core" - the cultural tourism reception area: it is the core function area of the Resort. The most important cultural tourism project is placed at the eastern gateway of the Resort - the central plaza area, to create an area with the most distinctive tourism characteristics and highlight the overall landscape of the Resort. To be specific, it includes high-end resort hotel complex on the mountain, Dai Show Theater, city-level business center - Wanda Plaza and Stylish Lakeshore Bar Street. With thorough exploitation into the local cultural characteristics and on the basis of featured cultural of the Dai Nationality, it is a region with the deepest cultural connotation and the most attractive charm in the Xishuangbanna International Resort.

(2) "Two Wings" - the east wing and the west wing: the New Tourism City lies at the east wing, focusing on the provision of ecological residence and supporting functions, among which public service functions, i.e. the Grade-A hospital, the primary school and secondary school, spread along the A1 main road to provide high-quality supporting services for the New Tourism City; the ecological communities with distinctive features locate in the eastern side of the A1 main road and are arranged along the loop with consideration to the landform. The Theme Entertainment District locates in the west wing, focusing on the Theme Park and sports, relaxation and entertainment. As the climax and cadenza of the Resort, this area locates in the west end of the central avenue and the Resort. As the destination of outdoor theme entertainment, this area is linked with the core areas via transportation to stimulate the circulation of visitors in the Resort.

4. REGIONAL CULTURE: CREATE A CULTURAL TOURISM AREA WITH CHINESE ELEMENTS, LOCAL CULTURAL AND PROMINENT ECOLOGICAL AND NATURAL CHARACTERISTICS

With rich tourism resources and known as an "ideal and magic paradise" in the ancient Dai language, Xishuangbanna is the one and only tropical rainforest natural reserve area with the unique resources and features of the Dai Nationality in China.

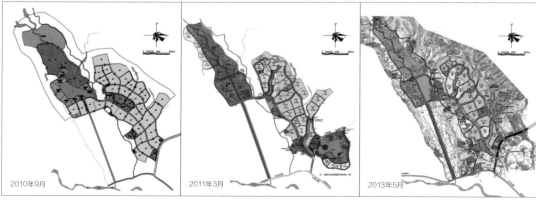
（图9）项目规划历程

廊道用地，构建度假区优越的生态环境，在有限制的条件下，创造富有特色的空间风貌和环境（图9）。

同时，更重要的是通过规划布局突出文旅，使文化旅游动线和布局符合消费者的需求，最大符合客观规律。度假区以道路山水为骨架，各功能区沿A1主路有序展开，公共功能及文化旅游业态布局在基地两端。具体而言，通过打造"一核"、"两翼"、"三带"，营造山水绿色之城。

（1）"一核"——文化旅游接待区：是度假区的核心功能区，将最重要的文化旅游项目放在度假区东侧门户——中央广场区，打造最具旅游度假特色的区域，突出展现了度假区的整体风貌。具体包括：山地高端度假酒店群、傣秀剧场、城市级商业中心"万达广场"及风情滨湖酒吧街。该片区充分挖掘地域文化特色，以傣族特色文化为底蕴，成为西双版纳国际度假区最具文化内涵、最具魅力的区域。

（2）"两翼"——东西两翼：东翼为旅游新城区，以生态居住和配套功能为主，其中"三甲"医院、中小学等公共服务功能沿A1主路展开，为旅游新城提供高品质的配套服务；生态社区在A1主路东侧区域，沿环路并结合地形布置各具特色的生态社区。西翼为主题娱乐区，以主题公园、体育休闲娱乐为主。该区是园区高潮及华彩部分，布局在中央大道和园区西侧尽端，作为室外主题娱乐目的地，通过交通与核心区联动起来，拉动园区人流。

（3）"三带"——冲沟景观带：主要依托三条现状冲沟的自然资源禀赋，对其进行改造升级，打造富有魅力的滨水休闲景观带。

4.地域文化——打造具有中国元素、地域文化及生态自然特点突出的文化旅游区

西双版纳旅游资源丰富，古傣语为"理想而神奇的乐土"，拥有中国唯一的热带雨林自然保护区和独特的傣族风情资源。

如果说迪士尼的成功凭借的是风靡全球的电影IP，那么万达在文化旅游项目上探索的是以区域文化为主题的新模式。万达西双版纳国际度假区应用高科技设备和手段，围绕热带雨林和傣族风情两个核心元素进行设计，精心打造了呈现原生态的雨林生活和傣族神话的"傣秀"，展现傣族建筑特色和南传佛教文化的山地度假酒店群，文化主题乐园等独具地域文化特色的文化旅游项目，吸引了全国各地、东南亚的旅游度假人群前来体验，成为西双版纳一张靓丽的"名片"。

度假区以景洪、西双版纳周边区域的森林、溪流、山涧和优越的气候、空气等自然生态环境为天然的生态基底环境，打造超大规模的公共生态绿带空间、地域文化特色及中国元素结合的多元建筑景观、统一而又丰富的生态地域景观元素，营造"山、水、城"相依的度假区整体景观意境，并坚持对度假区进行保护性开发、建设性利用，建设度假区完善的城市级生态景观绿地系统（图10）。

同时，度假区给水系统实现分质供水。景观供水采用澜沧江源水，践行绿色环保节能的水资源综合利用理念。

五、结语：万达西双版纳国际度假区——璀璨的明珠

万达西双版纳国际度假区作为万达集团第二代文化旅游产品，开创了高度融合传统地域文化的深度体验式旅游度假模式，将云南旅游及西双版纳城市发展水平提升到前所未有的高度，给云南旅游带来突破性的改变，是云南旅游转型升级的标志性项目及东南亚旅游新的目的地，成为澜沧江—湄公河流域、昆曼国际大通道旅游线区域一颗璀璨的明珠。

If the success of the Disney Land is attributed to its internationally welcomed movie IP, then the exploration on the cultural tourism projects made by Wanda is a new model that themed on the regional culture. With the application of high-tech equipment and techniques, the design of Wanda Xishuangbanna International Resort focuses on two core elements - tropical rainforest and the custom of the Dai Nationality. Meticulous efforts have been made to create cultural tourism projects with unique local cultural characteristics, i.e. the Dai Show that presents the original rainforest life and the mythology of the Dai nationality, the Mountain Resort Hotel Complex that shows the architectural features of the Dai nationality and the Southern Buddhism culture, and cultural theme park, which attracts visitors from all over the country and from Southeast Asia and makes it a showy attraction of the tourism in Xishuangbanna.

Taking the forests, brooks, mountain creeks in the surrounding areas of Xishuangbanna and its natural ecological environment (i.e. superior climate and air) as a natural ecological base environment, the public ecological greenbelt space of super-large scaled, multi-element construction landscape combining local cultural characteristics and Chinese elements and unified yet abundant ecological regional landscape elements are built in the Resort to create an overall artistic concept for the resort landscape of the coexistence of "Mountain, Water and City". In addition, the Resort sticks to the principle of protective development and constructive utilization to create a complete city-level ecological landscape green system inside the Resort (Fig. 10).

Meanwhile, the water supply system of the Resort adopts the principle of supplying water from different sources. The landscape water supply adopts the source water from the Lancang River to practice the green, environment-protection and energy-saving concept of comprehensive utilization of water resource.

V. CONCLUSION：WANDA XISHUANGBANNA INTERANTIONAL RESORT - A SHING PEARL

As one of the second generation cultural tourism products of Wanda Group, Wanda Xishuangbanna International Resort creates a tourism model that provides in-depth experience highly which merged with the regional culture. It elevates the development level of Yunnan's tourism industry andthe city of Xishuangbanna Yunnan's tourism industry to an unprecedented height, and brings ground-breaking changes for Yunnan's tourism industry. It becomes a landmark project for the transformation and upgrading of Yunnan's tourism industry and a new destination of tourism in Southeast Asia. It becomes a shining pearl in the tourism line of the Lancang-Mekong River Valley and the Kunming-Bangkok International Passage.

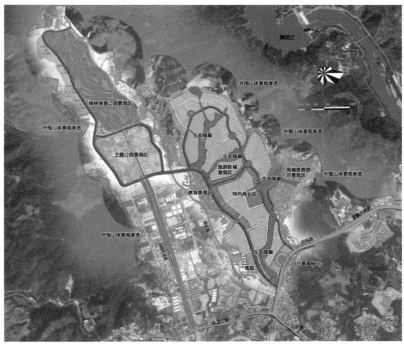

（图10）景观绿地分析图

02 PUBLIC LANDSCAPE OF WANDA XISHUANGBANNA INTERNATIONAL RESORT
万达西双版纳国际度假区公共景观

OVERVIEW OF PUBLIC LANDSCAPE
公共景观概况

万达西双版纳国际度假区占地5.3平方公里，园区内自然景观资源丰富，但大部分处于原始荒野的状态，不适合居住，需要通过公共景观的规划来满足宜居环境的需要。同时，公共景观把度假区内多个业态——主题公园、酒店群、住宅、酒吧街、大商业、配套公建等——串联起来，起到"融会贯通"的作用。

Covering a gross site area of 5.3 square kilometers, Wanda Xishuangbanna International Resort is rich in natural landscape resources. But most of the resources are in an original and wild state, which makes it unfit for residence. It is in need of public landscape planning to satisfy the demand for a livable environment. Meanwhile, public landscape links various business types (including Theme Park, hotel complex, residence, bar street, mega business and supporting public building) in the Resort together, fulfilling a function of bringing the business types together.

PLANNING CONCEPT
规划理念

西双版纳公共景观在设计之初即提出"尊重自然，保护环境"的宗旨，开发建设中充分考虑自然生态，并最终以"打造高品质的度假产品、提升度假区整体幸福感、改善当地居民生活环境为目的"。规划针对原始地形条件复杂、橡胶林带来的土壤贫瘠等现状条件加以充分调研分析后，对湿地公园的开发设计提出要求：第一，满足"功能第一"的原则；第二，满足充分尊重现状隐性条件、低成本开发及动态设计的原则。西双版纳公共景观将实现以上目标作为规划设计理念，使公共景观成为度假区内重要的"景观走廊"和"公共绿化空间"。

At the very beginning of designing the public landscape in Xishuangbanna, the purpose of "respecting nature and protecting environment" has been put forward. During the development and construction, the natural ecology shall be given full consideration to so as to realize the ultimate purpose of creating high-quality holiday product, lifting the overall happiness of the Resort and improving the living environment for local citizens. Based on a thorough research and analysis on the current conditions (including the complex original terrain conditions and the soil depletion brought by rubber plantation), the following requirements are raised against the development of wetland park: first, the principle of giving priority to function shall be satisfied; second, the principle of fully respect implicit current conditions, development at a low cost and dynamic design shall be satisfied. The design of public landscape in Xishuangbanna shall take the fulfillment of the above purposes as the planning and design philosophy to make public landscape the important "landscape corridor" and "public green space" in the Resort.

PLANNING MEASURES
规划手法

根据度假区整体规划的需要，通过"一路三带"的景观规划手法，实现了满足宜居环境的需要。通过公共景观有机地衔接串联起不同业态，在美化环境的同时又极大地提升了西双版纳国际度假区的品牌影响力，从而起到推广整个西双版纳项目的效果、担负起产业升级宣传的重任，成为西双版纳地区的视觉焦点。

Based on the overall planning requirement of the Resort, the demand for a livable environment is satisfied via the landscape planning technique of "One Road and Three Belts". Various business types are organically linked together by public landscape, which not only beautifies the environment but also promotes the brand influence of Xishuangbanna International Resort, thereby advancing the whole Xishuangbanna Project, shouldering heavy responsibility of industrial upgrading propaganda, and making it the visual focus of the Xishuangbanna area.

01 万达西双版纳国际度假区入口
02 万达西双版纳国际度假区景观大道

"ONE ROAD" - LANDSCAPE AVENUE OF WANDA XISHUANGBANNA INTERNATIONAL RESORT
"一路"——万达西双版纳国际度假区景观大道

规划时间：2012 – 2016	PLANNING PERIOD : 2012 - 2016
规划位置：云南 / 景洪	LOCATION : JINGHONG, YUNNAN PROVINCE
规划面积：全长3公里	PLANED AREA : 3KM
规划功能：景观大道、迎宾大道	FUNCTIONS : LANDSCAPE AVENUE, GUEST-MEETING AVENUE

PLANNING OVERVEIW
规划概况

景观大道起点度假区山门，是进入度假区的门户形象，全长近3公里，根据周边的业态特征分为"十二吉象"、"浪漫花海"、"林荫大道"和"热带风情"四大主题；以澜沧江文化为主线，将丰富的植被、孔雀、大象等当地特色元素集一身，突出提炼了西双版纳的人文风情，欢迎五湖四海的朋友光临万达西双版纳国际度假区。

With its origin starting from the main entrance of Xishuangbanna International Resort and an overall length of about 3km, the landscape avenue serves as a gateway image of the Resort. It is divided into four themes of "Twelve Auspicious Elephant", "Romantic Flower Sea", "Park Avenue" and "Tropical Custom" as per the characteristics of surrounding business types; taking the Lancang River culture as a mean line, abundant elements of the local characteristics (i.e. plants, peacock and elephant) are combined into one to extract the cultural customs of Xishuangbanna, attracting people from all corners of the country to visit Wanda Xishuangbanna International Resort.

03
03 万达西双版纳国际度假区沿河景观
04 万达西双版纳国际度假区商业街景观雕塑
05 万达西双版纳国际度假区湿地公园总平面图
06 万达西双版纳国际度假区湿地公园沿河景观
07 万达西双版纳国际度假区湿地公园绿化

PLANNING ACHIEVEMENT
规划成果

山门景观及道路景观充分表现了傣族文化，配以丰富多样的热带雨林植物、珍贵树种、异彩缤纷的各色花卉，形成了度假区内独具特色的湿地雨林景观空间；作为特色鲜明的度假区门户形象，重塑了度假区的整体度假环境并提升了度假区的品质及氛围。

The entrance landscape and road landscape sufficiently represent the culture of the Dai Nationality. A wetland rainforest landscape space of unique features has been formed with diversified tropical rainforest plants, rare trees and radiant and colorful flowers; as the gateway image with distinctive features of the Resort, it remodels the overall holiday environment of the Resort and promotes the quality and atmosphere of the Resort.

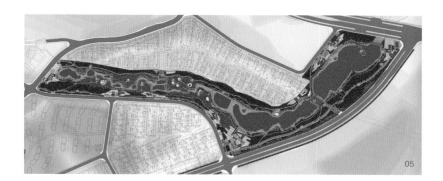

"THREE BELTS" - WETLAND PARK OF WANDA XISHUANGBANNA INTERNATIONAL RESORT
"三带"—万达西双版纳国际度假区湿地公园

规划时间：2012 / 10 - 2016 / 09
规划地点：云南 / 景洪
规划面积：10公顷
规划功能：湿地公园，休闲、慢跑健身公园

PLANNING PERIOD：OCTOBER 2012 TO SEPTEMBER 2016
LOCATION：JINGHONG, YUNNAN PROVINCE
PLANED AREA：10 HECTARES
FUNCTIONS：WETLAND PARK, RECREATION PARK, JOGGING FITNESS PARK

PLANNING OVERVIEW
规划概况

西双版纳国际度假区规划"三大湿地公园"，总面积34公顷，约占度假区总景观面积的38%，如同"三块绿色宝石"提高了度假区居住品质。景观设计利用湿地本身丰富的跌水及本地特色水生植物，通过层层过滤净化把湿地公园的现状水体得以合理优化，使湿地公园内原有的水质从Ⅲ类提升到Ⅱ类，极大地提高了园区的整体环境质量。

With an overall site area of 34 hectares and accounting for 38% of the overall landscape area in the Resort, the "Three Wetland Parks" in Xishuangbanna International Resort promote the residence quality of the Resort like three pieces of green gemstones. Taking advantages of the abundant cascading waterfalls of the wetland and aquatic plants of the local characteristics, the landscape design reasonably optimizes the current water system of the wetland park via multi-layered filtration and purification, thereby improving the water quality of the Wetland Park from the original Grade III to Grade II and significantly improves the overall environment quality of the Resort.

08 万达西双版纳国际度假区湿地公园景观
09 万达西双版纳国际度假区湿地公园景观栈桥
10 万达西双版纳国际度假区湿地公园水景
11 万达西双版纳国际度假区湿地公园景观栈桥

PLANNING ACHIEVEMENT
规划成果

第一，"1号湿地公园"在解决水利泄洪的基本功能下，对水质的净化体系带来了惊人成效。水系从湿地公园的最北端到最南端，经过五级的跌水，水质的净化成果非常显著，并且经检验无一处渗漏。第二，"1号湿地公园"取得突出的景观效果，成为一处景色优美、具有西双版纳当地地域特色，集休闲、慢跑、健身等功能于一体的多功能湿地景观公园。第三，"1号湿地公园"成为联系住宅、秀场、酒吧街等业态的绿色纽带。

First, in addition to the basic functions of solving water conservation and flood discharge problems, the "No.1 Wetland Park" also brings amazing effect to the water quality purification system. After five layers of cascading waterfall from the northernmost end of the Wetland Park to the southernmost end, the result of water quality purification is quite remarkable, and upon testing, there is no sign of leakage. Second, the "No.1 Wetland Park" achieves outstanding landscape effect, and becomes a multi-functional wetland landscape park with beautiful scenery, local regional features of Xishuangbanna and functions of relaxation, jogging and bodybuilding. Third, the "No.1 Wetland Park" becomes a green belt that links various business types including residence, show theater and bar street.

03 HOTEL COMPLEX OF WANDA XISHUANGBANNA INTERNATIONAL RESORT
万达西双版纳国际度假区酒店群

OVERVIEW OF THE HOTEL COMPLEX
酒店群概况

2015年9月26日,万达西双版纳国际度假区酒店群盛大开业,三座豪华酒店——西双版纳万达文华度假酒店、西双版纳万达希尔顿逸林度假酒店、西双版纳万达假日度假酒店——各具特色、震撼亮相。酒店群总用地46.67公顷(700亩),共980间客房,景观面积超过43万平方米,拥有9800余棵热带树木,场地内的植物资源几乎涵盖了东南亚所有常见的热带植物品种。

On September 26, 2015, Wanda Xishuangbanna International Resort Hotel Complex started business with grand debut of its three luxury hotels - WANDA VISTA RESORT XISHUANGBANNA, DOUBLETREE RESORT XISHUANGBANNA BY HILTON HOTEL and CROWNE PLAZA RESORT XISHUANGBANNA PARKVIEW, which have distinctive features from each other. With an overall site area of 46.67 hectares (700 mu), 980 guestrooms, 9,800 tropical trees and the landscape area being over 430,000 square meters, the plant resources in the site almost cover the whole species of tropical plants in Southeast Asia.

01 万达西双版纳国际度假区酒店群效果图
02 万达西双版纳国际度假区酒店群鸟瞰图

02

OVERALL PLANNING
整体规划

酒店群总体规划遵循"融于自然、绿色生态、山地度假、文化体验"的原则。为了充分发掘西双版纳"理想而神奇的天堂乐土"之美，在酒店规划动笔之前，主创设计及管理团队到泰国清迈、苏梅岛等地考察了十几家顶级度假酒店，在充分理解度假酒店特点的基础上，确立了将酒店融入风景的策划理念。

酒店规划设计着力于在旅居领域寻找与世界同步的本土价值，以"场所精神"为研究重点，秉持"传统新体验"为设计理念，深入挖掘所在地的地域文化精神，将其高度融合并以全新设计方式打造"创新体验"，创意地打造这片"隐秘而浪漫"的世界顶级度假胜地。

在布局上，三座酒店布局顺应地形、随山就势、依次展开。整体建筑布局最大限度地利用地形地势和冲沟的位置特点，将道路设于基地西侧，在较低处设置两座五星酒店，使其面向东侧葱郁的山地；而在地块最高处布置了六星文华酒店，大堂、餐厅、公共泳池等公共区域沿水景轴线转而面向西侧绵延的群山，正对酒店大堂轴线西端设置了西双版纳总佛寺——禅修堂佛塔。酒店建筑、佛塔与群山的倒影融为一体，获得了"天人合一"、静谧超然的意境。

六星酒店的联排泳池别墅依山梯次而建，上层均能看到南方远山的风景。每套客房均配备热带花园、宽敞的露台和私家户外泳池，沉浸其中，心旷神怡。于东侧隐秘的山谷丛林中，依坝成湖修建的SPA区独居一隅，成为整个酒店群的最隐秘幽静的区域——丛林山谷与湖光倒影之中，创造了一处休闲养生的极致仙境。

酒店群的规划设计充分发掘了西双版纳热带雨林自然环境和少数民族风情文化的独特资源，通过对场地的山地环境、热带雨林气候和浓郁的傣族风情文化的研究和总结，尊重当地环境特色，以当地文脉串联酒店度假功能，呈现出融合于自然环境的气质，形成鲜明文化氛围的整体规划理念，提供了一个隐于自然风景、私密幽静的度假胜境。

The overall planning of the hotel complex follows the principle of "Nature Immersion, Green Ecology, Resort in Mountainous Region and Cultural Experience". In order to fully explore the beauty of Xishuangbanna as an "Ideal and Magic Paradise", before starting the planning of the hotel, the main design and management team has inspected more than ten top-level resorts in Chiengmai and Koh Samui, Bangkok, and on the basis of a thorough understanding of the features of resort hotels, the planning concept of merging hotel into scenery has been settled.

The planning and design of the hotel complex strives to find out values of native arts that are synchronized with the world in the field of tourism and residence. Taking the "Place Spirit" as the focus of research and adhering to the design concept of the "traditional new experience", the design probes deeply into the regional cultural spirit of the project site, and upon a high degree of integration, creates "innovative experience" with brand new design methods, thereby building a "covert and romantic" world-class top resort.

With the overall layout conforming to landform, the three hotels spread in sequence along the mountains. The overall architectural layout makes the most of the features of the landform, terrain and gully location. The road is set in the west side of the site, and two five-star hotels are set at the lower spot, facing the verdant mountains in the east side; while, the six-star Vista Hotel is set at the highest sport, with the public areas (i.e. the hall, restaurant and public pool) facing the stretching mountains in the west side along with the waterscape axis. Xishuangbanna Head Buddhist Temple, the Meditation Hall Pagoda, is set directly at the west side of the axis of the hotel hall. With an integration of hotel buildings, the pagoda and the reflection of mountains into one, a quiet and transcendent sight with "the unity of human and nature" is acquired.

The townhouses with swimming pools of the six-star hotel are built in sequence along the mountain. Visitors can enjoy a view of the remote mountains in the south from the upper storey. Every guestroom is equipped with a tropical garden, spacious terrace and private outdoor pool, rendering visitors a relaxed and happy experience by immersing in it. Located in the covert valley and jungle in the east side, the SPA area built along the dam and lake occupies an independent corner, becoming the most secluded and peaceful area in the whole hotel complex - a perfect fairyland for relaxation and health maintenance is created among the jungle, the valley and the reflection of lake view.

The planning and design of the hotel complex fully explores the unique resources of the tropical rainforest natural environment and the folk-custom of the ethnic group in Xishuangbanna. Based on a research and conclusion of local mountain environment, tropical rainforest climate and distinctive customs and culture of the Dai Nationality and with respect for local environment features, the holiday functions of the hotel are linked together via local design and context, presenting a disposition with the fusion of natural environment, forming an overall planning concept with distinctive cultural atmosphere and providing a sacred, private and peaceful site for vacation that is hidden in the natural scenery.

SPATIAL CREATION
空间创造

万达西双版纳国际度假区酒店群整体规划设计充分挖掘了热带雨林自然环境与少数民族文化风情的特征，以西双版纳傣族传统村落为原型，随山采形，并融入现代规划设计理念，全新打造空间环境；结合山地形成迂回的入口和丰富的台地，构建具有独特佛塔气场与异域风情生活方式相结合的隐秘王宫，创造出独特的"民族、自然、文化"相结合的建筑空间之美。山脚处的酒店群山门界定了度假的空间范围，沿着林荫道路可到达西双版纳万达希尔顿逸林度假酒店和西双版纳万达假日度假酒店。作为大型度假酒店，这两座酒店空间开放，设施齐全，满足家庭化度假需求。此区域的空间营造讲究"大开大阖"，错落有致，给客人以丰富的体验并满足多种需求。沿路继续上行，六星酒店山门及前院起到了"欲扬先抑"的作用，到达景观桥则空间豁然开朗，六星文华酒店整体展现在面前。文华酒店的公区空间宜人亲切，强调客人的尊贵体验；而文华酒店的客房区则沿山势蜿蜒上行，层层叠叠，给客人以傣寨村落的体验；SPA区更是利用山谷的湖水形成静谧的雨林空间。

酒店群空间结构上秉承山地形态，结合不同酒店功能设置，形成由山下至山顶，由开放至隐秘的空间氛围。酒店群以傣寨的模式，依山梯级而建，公共区域和客房院落及别墅的院落层次均很好地体现了聚落的精神。公共区域是聚落的中心，通过院落、景观、水池等梯级向西山方向层叠展开，文华酒店西侧的佛塔成了其公区轴线的中心焦点。联排别墅客房利用山势地形的高差，上下两排错落组合——上层客房从北侧坡上的前院进入，而下层客房从南侧坡下的泳池花园进入；下层别墅的屋顶成了上层别墅的泳池花园，上下层次的客房与前后院落的错动相得益彰。总统别墅和套房别墅置于山势的高处，大小别墅的排列也反映了聚落尊长的层级关系。

The overall planning and design of Wanda Xishuangbanna International Resort Hotel Complex explores deeply into the characteristics of the natural environment of tropical rainforest and the ethnic group culture and custom. Taking the Xishuangbanna Dai traditional village as the prototype and forming shape along the mountain, modern planning and design concept is adopted in to create a new spatial environment; circuitous entrances and abundant platforms are formed in combination with the mountains. A secrete palace combined with distinctive pagoda aura and exotic life style is built to create the unique beauty of architectural space that combines nationality, nature and culture. The mountain gate at the foot of the mountain defines the spatial scale of the vacation. The Doubletree Resort by Hilton Hotel Xishuangbanna and Crowne Plaza Resort Xishuangbanna Parkview can be accessed via the alameda. As large-scale resorts, these two hotels have open spaces and complete facilities to satisfy the vacation demand of families. The space construction of this region is particular about "large-scale spatial separation" with well-arranged areas to provide guests with abundant experience and satisfies various needs of the guests. If one keeps on walking along the road, he would see the gate of the six-star hotel at first, which fulfills the function of foreshadowing. After he reaches the landscape bridge, the space is suddenly enlightened with the full image of the six-star vista hotel revealing in front of him. The space of public area of Wanda Vista is pleasant and cordial, highlighting the exalted experience of the guests; while, the guestroom area of Wanda Vista Resort winds along the mountain tier upon tier, providing guest with the experience of the Dai village; what's more, the SPA area utilizes the lake water in the valley to form a quiet rainforest space.

The spatial structure of the hotel complex conforms to the landform of the mountain. In combination with different hotel function sets, a space atmosphere from the top of the mountain to the foot of the mountain, from openness to secrecy is formed. The hotel complex, taking the model of the Dai village, is built in sequence along the mountain. The gradation of public area, the guestroom courtyard and the villa all greatly reflect the spirit of human settlement. The public area, as the center of the settlement, spreads layer upon layer towards the direction of the western mountain via the layer of courtyard, landscape and water pool. The pagoda to the west side of the Wanda Vista Resort becomes the central focus on the axis of the public area. Taking advantage of the altitude difference of the mountains and landform, the townhouse guestroom area forms a scattered combination of upper row and lower row - the guestroom in the upper row can be accessed from the forecourt on the northern slope, while the guestroom in the lower row can be accessed from the pool garden in the southern row; the roof of the lower row villa becomes the pool garden of the upper row villa, and the scattered front and back courtyard bring out the best in each other. The president villa and suite locate in the upper spot of the mountain, and the arrangement of the big and small villa also shows the hierarchical relationship among the elders and betters of the settlement.

CREATION OF ARTISIC CONCEPTION
意境营造

万达西双版纳国际度假区酒店群通过不同的文化主题区将不同的功能空间融为一体，创造出独具文化及自然特色的空间意境。酒店群整体设计上通过形象明确的地域文化元素，以及典型的当地材料和建造工艺，在不同的功能区域营造出对版纳地域文化的不同体验，以"让客人发现故事"的方式来放慢身心节奏，体验到不同酒店独有的傣式度假酒店文化气质和轻松的度假氛围。

整个酒店群的空间随着星级的不断提升，形成由公共到私密、自然的体验，最终达到与自然的完美融合。依山而建的六星文华酒店如同掩映在雨林绿树中的隐秘王宫；版纳"传统傣族文化主题"、禅修堂的"南传佛教禅修文化主题"和文华酒店SPA区的"热带雨林文化主题"都通过不同的建筑处理、景观元素和植物配置来营造出不同的空间意境，让建筑从内外空间上的塑造延伸至精神层面，给客人以独特的心灵体验。这里处处尊贵典雅又亲近自然，可以同时体验鲜明的异域风情生活方式和浓厚的当地宗教文化氛围，配以傣族特色的酒店设施与顶级的服务，而获得无与伦比的度假感受。

万达西双版纳国际度假区酒店群利用灵动流畅的空间设计，将建筑与自然完美融合在一起，"步移景异"之间，打造出独具特色的山地热带雨林空间之美。酒店力求凸显独特的地域文化特色，延续区域文脉，将当地傣族的生活方式与宗教文化元素融入设计，带给客人清新独特的感受，结合优雅的环境与顶级的服务，构建一个在热带雨林的生态环境中放松身心、净化心灵的场所，为休闲度假的客人营造一个舒适、精致、尊贵的顶级度假天堂。

Wanda Xishuangbanna International Resort Hotel Complex merges different function spaces via different cultural theme area to create a space concept of unique cultural and natural features. With the adoption of the local cultural elements of explicit image and typical local materials and building techniques, the overall design of the hotel creates different experience of Xishuangbanna local culture at different function areas, enabling guests to slow down the pace of their mind and body by means of "discovering stories" and experience the unique cultural ethos of the Dai-style resort and relaxed holiday atmosphere of different hotels.

Along with the elevation of star-level, the space of the whole hotel complex forms an experience changing from public to secret & nature, and ultimately the perfect fusion with nature has been achieved. The six-star Wanda Vista Resort built along the mountain is like a secret palace hidden in the green trees of the rainforest; different space concepts are formed for the Xishuangbanna "Traditional Dai Nationality Cultural Theme", the "Southern Buddhism Meditation Cultural Theme" of the meditation hall and the "Tropical Rainforest Cultural Theme" of the SPA area of Wanda Vista Resort Hotel via different architecture treatment and the arrangement of landscape elements and plants, extending the modeling of building from the inner and outer space to spiritual level and providing guests with unique spiritual experience. This place is not only noble, elegant and close to nature at every spot, but also gives people an opportunity to experience an exotic life style and strong local religious cultural atmosphere. Matched with hotel facilities of the features of the Dai Nationality and top-level services, guests can enjoy an unrivalled holiday experience.

Wanda Xishuangbanna International Resort Hotel Complex perfectly merges architecture with nature via flexible and smooth spatial design, and creates the distinctive beauty of mountain tropical rainforest space by varying sceneries with changing view points. The hotel strives to highlight the distinctive local cultural features, extend local design and context and bring guest a fresh and unique experience with the fusion of local life style of the Dai Nationality and religious cultural elements. In combination with elegant environment and top-level service, a place for the relaxation of mind and body and the purification of spirit is created in the ecological environment of tropical rainforest, offering a comfortable, delicate and exalted top-level paradise for vacation for guest on a relaxing vacation.

PART **B** | WANDA XISHUANGBANNA INTERNATIONAL RESORT 万达西双版纳国际度假区 | 035

WANDA VISTA RESORT XISHUANGBANNA
西双版纳万达文华度假酒店

时间：2015 / 09 / 26　　OPENED ON : 26th SEPTEMBER, 2015
地点：云南 / 景洪　　　LOCATION : JINGHONG, YUNNAN PROVINCE
建筑面积：4.65万平方米　FLOOR AREA : 46,500M²

PROJECT OVERVIEW
酒店概况

西双版纳万达文华度假酒店占地16.0公顷，总建筑面积4.65万平方米，拥有客房151间。

Wanda Vista Resort Xishuangbanna covers an area of 16.0 hectares, has a gross floor area of 46,500 square meters, and accommodates 151 rooms.

03 万达西双版纳国际度假区酒店群总平面图
04 西双版纳万达文华度假酒店公共区内庭入口

DESIGN CONCEPT
设计理念

西双版纳万达文华度假酒店秉承"将酒店融入风景"的策划理念，整体坐落在山林的优美环境中，建筑融于自然，也成为风景的一部分。

文华酒店将酒店合理规划整体功能分区、分体量布置，注重围合和开敞的有机结合、造景与借景的有机结合，将大堂、餐厅、公共泳池等公区沿水景轴线向西山方向布置，获得静谧超然的意境；而客房联排别墅依山梯次而建，均能看到南方远山的风景；雨林中围绕人工湖的SPA别墅群，更增加了林中仙境的感受。

文华酒店的客房区全部采用联排别墅式设计。客房利用山势地形的高差，分上下两排错落组合，上层的客人可以从北侧坡上的前院进入客房，而下层的客人则从南侧坡下的泳池花园进入客房；上层别墅的泳池花园盖在下层别墅的屋顶上；上下两层的客房与前后院落的错动相得益彰。

在酒店东侧的丛林中，利用冲沟地势拦坝成湖，五栋SPA别墅依湖而建，从酒店客房区到SPA湖经丛林中小路蜿蜒而至。客人置身湖光山色，享受置身林中仙湖的意境。

万达文华度假酒店是世界上首个以南传佛教佛塔、禅修堂为中心而建的顶级度假酒店。佛塔采用傣族特有的造型形式，成为地域独特文化内涵的中心。周边遍布与南传佛教相关的"五树六花"，行走其间，隐褪浮躁，杂念尽消。酒店整体景观以"佛

Adhering to the planning concept of "merging hotel into scenery", Wanda Vista Resort Xishuangbanna locates in the beautiful environment of mountains and forests, with the building merged into nature and became a part of the scenery.

Wanda Vista Resort adopts an overall reasonable planning with function division and arrangement as per construction volume. The design pays attention to the organic combination of enclosure and openness, as well as the combination of landscaping and view borrowing. The public areas (i.e. hall, restaurant and public pool) are arranged towards the direction of mountains in the west side along the axis of waterscape to acquire a prospect of tranquility and transcendence; while, the townhouses (guestrooms) are built in sequence along the mountain with a view of the remote mountains in the south; the SPA villa complex surrounding the man-made lake in the rainforest further enhances the experience of fairyland in the forest.

The guestroom area of Wanda Vista Resort Xishuangbanna completely adopts the design of townhouse. Taking advantage of the altitude differences of the mountain landform, the guestrooms adopt a scattered upper and lower two-row arrangement. Guests on the upper row can enter into the guestroom via the forecourt in the northern slope, while guests on the lower row can enter into the guestroom via the pool garden at the foot of the southern slope; the pool garden of the upper villa is built on the roof of the lower villa; the upper and lower two-storied guestroom, and the scattered front and back courtyard bring out the best in each other.

In the jungle at the east side of the hotel, a lake is formed with dams constructed in accordance with the landform of gullies. Five SPA villas are built around the lake, which can be accessed from the guestroom area of the hotel via the winding path in the jungle. Exposed in the landscape of lakes and mountains, guests are able to enjoy the prospect of placing themselves in a fairy lake inside the forest.

Wanda Vista Resort Xishuangbanna is the first top-level resort centered on Southern Buddhism Pagoda and meditation hall. The pagoda adopts design forms that are unique of the Dai Nationality and becomes the center of local-unique cultural connotation. "Five Trees and Six Flowers" related to Southern spread all over the surrounding areas. People would be free from restless mind and distracting thoughts by walking among them. The overall landscape of the hotel adopts "Buddhism and Zen" as the theme. From the moment that the guest enters into the hotel located in the mountains and dense forest, he starts to experience an elaborately designed and interlocking plot - a premier and covert breath and scene is shaped with the life style, religious, and cultural atmosphere of the features of the Dai Nationality,

05 西双版纳万达文华度假酒店公共区游泳池
06 西双版纳万达文华度假酒店入口连廊
07 西双版纳万达文华度假酒店全日餐厅水景面

意禅修"为主题，客人从进入群山密林中的酒店开始，就在体验设计铺陈的环环相扣情节——傣族特色的生活方式和宗教文化氛围，塑造出尊尚而隐秘的气息与场景，使参与和融入的客人获得前所未有的异域情境体验。酒店东侧设有一处僧舍禅院，与佛塔一起构成完整的西双版纳总佛寺禅修堂功能。在茂林之间咏禅颂佛，为静谧的雨林酒店带来了超然的气氛。孔雀、瑞象、莲花，都是傣族最具代表性的图腾标志和宗教符号，也是傣族民族精神的象征。酒店建筑装饰细节及各处的景墙、铺装，充分体现出这种独特的以自然生物与文化符号形成的主题元素。

开业一年以来，西双版纳万达文华度假酒店屡获国际、国内大奖，被意大利著名设计行业网站Designboom评选为2015年"全球十佳度假酒店"；作为中国唯一酒店同时入围"第十二届国际HD Awards"（Hospitality Design Awards）之"最佳度假酒店奖"和"最奢华酒店公区奖"两项大奖；荣获2016"美丽中国金橄榄奖·中国最佳景观设计酒店"大奖等。正如评委们在获奖理由中表述的一样：西双版纳万达文华度假酒店的获奖，源于酒店以西双版纳地域文化性多种形式介入，贯穿丰富的主题文化内涵，带给酒店入住者有价值的体验。它的设计成功地制造了场所的独创性、灵活性与文化性，作品的美感、品质和内涵的完美统一。

rendering an unprecedented experience of foreign scenario to guests that participate and blend in. A Buddhist hall is built in the east side of the hotel and forms the complete function of Xishuangbanna Head Pagoda Meditation Hall together with the pagoda. Chanting Buddhist scriptures among the dense forest brings a transcendent atmosphere to the rainforest hotel. Peacock, elephant and lotus are all most representative totem and religious marks of the Dai Nationality, as well as the token of the spirit of the Dai Nationality. The architectural decoration details and the landscape wall and pavement of the hotel adequately represent such unique theme element formed by natural biology and cultural symbol.

In the past one year since its opening, Wanda Vista Resort Xishuangbanna has won several international and domestic awards, and is awarded as "Top Ten Resorts in the World" by Designboom, a renowned Italian Design Website; it is also the only Chinese hotel that enters the final lists of "Best Resort Hotel Award" and "Luxurious Hotel Public Area Award" of the Twelfth International Hospitality Design Awards; it wins the 2016 "Beautiful China Gold Olives Award - Chinese Hotel with the Best Landscape Design". As stated by the jurors in the justification for the award: the reason why Wanda Vista Resort Xishuangbanna wins the award lies in the value experience that it delivers to its customers via the introduction of various forms with regional culture of Xishuangbanna and the penetration of rich theme cultural connotation. Its design successfully creates the originality, flexibility and culture of the site, achieving a perfect unification of aesthetic feeling, quality and connotation of the works.

INTERIOR DESIGN
酒店内装

ECOLOGY LANDSCAPE
生态景观

酒店景观设计突破常规设计手法，致力融合自然环境，引入当地独特的地域文化元素，以西双版纳特有的建造工艺和材料，营造出别样的文化体验模式。让客人以"发现故事"的方式来放松身心，体现出本酒店独有的精神气质和轻松的度假氛围。基于上述原则，结合西双版纳文华酒店的功能分布，通过"三个文化主题"，分别是：以象征意义的植物（"五树六花"）配置为主打造的佛塔区，以孔雀、大象、莲花、守护神等傣式传统文化元素为主的版纳民族文化主题区以及雨林原生态文化主题的SPA区——共同打造出了热带雨林度假酒店景观。

The landscape design of the hotel breaks through the conventional design methods with an effort to blend in natural environment and bring in regional cultural elements of unique local features. A peculiar cultural experience model is created with the construction techniques and materials that are unique in Xishuangbanna. The hotel aims to let the visitors relax themselves by means of "discovering stories", which represents the unique spiritual temperament and relaxing holiday atmosphere of the hotel. Based on the above principle and in consideration of the function distribution of Wanda Vista Resort Xishuangbanna, a tropical rainforest resort landscape is created with the "three cultural themes", which are the pagoda area built with plants of symbolic meanings ("Five Trees and Six Flowers"), the Xishuangbanna ethnic culture theme park with the traditional Dai cultural elements (i.e. peacock, elephant, lotus and patron saint), and the SPA area with the theme of rainforest original culture.

10

08 西双版纳万达文华度假酒店佛塔
09 西双版纳万达文华度假酒店大堂
10 西双版纳万达文华度假酒店SPA入口
11 西双版纳万达文华度假酒店内庭

11

NIGHTSCAPE DESIGN
酒店夜景

12 西双版纳万达文华度假酒店观景平台
13 西双版纳万达文华度假酒店落客区
14 西双版纳万达文华度假酒店全日餐厅夜景

DOUBLETREE RESORT BY HILTON HOTEL XISHUANGBANNA
西双版纳万达希尔顿逸林度假酒店

时间：2015 / 09 / 26　　**OPENED ON**：26th SEPTEMBER, 2015
地点：云南 / 景洪　　　**LOCATION**：JINGHONG, YUNNAN PROVINCE
建筑面积：4.56万平方米　**FLOOR AREA**：45,600M²

PROJECT OVERVIEW
酒店概况

西双版纳万达希尔顿逸林度假酒店占地5.96万平方米，总建筑面积4.56万平方米，拥有房间413间。

Doubletree Resort by Hilton Hotel Xishuangbanna has a site area of 59,600 square meters, a gross floor area of 45,600 square meters, and accommodates 413 rooms.

15 万达西双版纳国际度假区酒店群总平面图
16 西双版纳万达希尔顿逸林度假酒店客房鸟瞰图

17 西双版纳万达希尔顿逸林度假酒店客房
18 西双版纳万达希尔顿逸林度假酒店大堂吧夜景
19 西双版纳万达希尔顿逸林度假酒店入口夜景
20 西双版纳万达希尔顿逸林度假酒店入口
21 西双版纳万达希尔顿逸林度假酒店大堂

DESIGN CONCEPT
设计理念

西双版纳万达希尔顿逸林度假酒店公共区以傣族民居为主题，利用傣族高耸的屋顶为基本元素，辅以当地的砂岩、瓦、浮雕等材料，利用现代的设计手法，勾勒出高低错落、曲折起伏的天际线及宜人尺度的院落空间，体现出建筑与当地人文环境相契合的度假风格。

With the theme of dwellings of the Dai Nationality, the design of Doubletree Resort by Hilton Hotel Xishuangbanna adopts soaring roof of the Dai Nationality as the basic element, accompanied by materials like sandstone, tile and relief of the day. Scattered and circuitous skyline and courtyard space with favorable size are formed with modern design techniques, revealing a holiday style of the building that agrees with the local cultural environment.

INTERIOR DESIGN
酒店内装

ECOLOGY LANDSCAPE
生态景观

西双版纳万达希尔顿逸林度假酒店景观采用多种类植物及景观小品，大量运用榕树类、蕉类、雨树等当地苗木品种，配合打造出具有浓郁地域风情的热带雨林景观。整个酒店景观绿化面积约3.1万平方米，拥有1100余棵乔木，使整个酒店掩映在植物之中，营造出了傣式度假酒店风格。

Various kinds of plants and landscape accessories are adopted to create the landscape of Doubletree Resort by Hilton Hotel Xishuangbanna. Local nursery stocks (including banian, musa and saman) are extensively used to create a tropical rainforest landscape with strong local characteristics. With an overall landscape green area of about 31,000 square meters and about 1,100 trees, the whole hotel is covered with plants and is rendered with a style of the Dai Nationality.

22-24 西双版纳万达希尔顿逸林度假酒店内庭

CROWNE PLAZA RESORT XISHUANGBANNA PARKVIEW
西双版纳万达皇冠假日度假酒店

时间：2015 / 09 / 26
地点：云南 / 景洪
建筑面积：4.6万平方米

OPENED ON：26th SEPTEMBER, 2015
LOCATION：JINGHONG, YUNNAN PROVINCE
FLOOR AREA：46,000M²

PROJECT OVERVIEW
酒店概况

西双版纳万达皇冠假日酒店占地6.2万平方米，总建筑面积4.6万平方米，拥有房间416间。

Crowne Plaza Resort Xishuangbanna Parkview has an area of 62,000 square meters, an area of 46,000 square meters and accommodates 416 rooms.

25 西双版纳万达皇冠假日度假酒店全景图
26 万达西双版纳国际度假区酒店群总平面图

DESIGN CONCEPT
设计理念

西双版纳万达皇冠假日酒店结合傣民族特有的元素，在酒店落客区、大堂、宴会厅等位置设计气势恢宏的坡屋面。酒店立面随地形高差逐级错落，塑造出层次分明且舒展大气的建筑形象。砂岩、涂料、木构架色彩鲜明又浑然一体，远远望去，整个酒店与山地的林木交相呼应，彼此共生，体现出度假酒店低调奢华、亲近自然的地域文化属性。

Adopting the unique elements of the Dai Nationality, imposing pitched roofs are designed in locations like drop-off area, hall and banquet hall of the Crowne Plaza Resort Xishuangbanna Parkview. The elevation of the hotel is scattered along with the altitude differences of the landform, creating an architectural image of distinct gradation and stretched magnificence. The sandstone, painting and wood structures are bright in color and blend into the surrounding environment. From a distance, it seems that the hotel echoes and co-exists with the woods in the mountain, revealing the regional culture attributes of modest luxury and love for nature of the Resort.

INTERIOR DESIGN
酒店内装

27 西双版纳万达皇冠假日度假酒店落客区
28 西双版纳万达皇冠假日度假酒店大堂
29 西双版纳万达皇冠假日度假酒店康体区全景图

ECOLOGY LANDSCAPE
生态景观

西双版纳万达皇冠假日酒店景观采用多种类植物及景观小品，大量运用榕树类、蕉类、雨树等当地苗木品种，配合打造出具有浓郁地域风情的热带雨林景观，体现傣式度假酒店风格。整个酒店景观绿化面积约3万平方米，拥有1500余棵乔木。场地内的植物资源几乎涵盖了东南亚所有的热带植物品种，客人在享受这座热带植物天堂的同时，也增长了知识、愉悦了身心。

Various kinds of plants and landscape accessories are adopted to create the landscape of Crowne Plaza Resort Xishuangbanna Parkview. Local nursery stocks (including banian, musa and saman) are extensively used to create a tropical rainforest landscape with strong local characteristics and reveal a style of the Dai Nationality. The overall landscape green area of the hotel amounts to about 30,000 square meters with about 1,500 trees. The plant resources in the site almost cover the whole variety of tropical plants in Southeast Asia. Visitors can learn knowledge as well as relax their mind and body while they enjoy the paradise of tropical plants.

30 西双版纳万达皇冠假日度假酒店康体区外立面
31 西双版纳万达皇冠假日度假酒店全日餐厅外景观
32 西双版纳万达皇冠假日度假酒店主景观池

04 BAR STREET OF WANDA XISHUANGBANNA INTERNATIONAL RESORT
万达西双版纳国际度假区酒吧街

规划时间：2012 / 10 – 2016 / 09	**PLANNING PERIOD**: OCTOBER 2012 TO SEPTEMBER 2016
规划地点：云南 / 景洪	**LOCATION**: JINGHONG, YUNNAN PROVINCE
规划面积：10 公顷	**PLANNED AREA**: 10 HECTARES
建筑面积：15818 平方米	**FLOOR AREA**: 15,818M²
规划功能：酒吧、休闲、娱乐	**FUNCTIONS**: BAR, RECREATION, ENTERTAINMENT

01

PROJECT OVERVIEW
规划概述

酒吧街位于湖边，总建筑面积为10863平方米，共计103间店铺。酒吧街沿进入度假区的主路和湿地公园次第展开：一面临街，交通便利，繁华聚集；一面临湖，眺山亲水，环境优雅。设计因地制宜，就水取势，形成极具魅力的风情休闲街区。

With an overall floor area of 10,863 square meters and 103 shops, the bar street locates by the lakeside. The bar street spreads in sequence along the main road and the wetland park of the Resort: one side of it faces the street, enjoying convenient transportation and prosperity; the other side faces the lake, enjoying with a view of the remote mountain and the nearby water, and an elegant environment. Designed according to local landform and the flow direction of the water, a recreation street of unique charm is formed.

02

01 万达西双版纳国际度假区酒吧街全景图
02 万达西双版纳国际度假区酒吧街建筑外立面

DESIGN CONCEPT
设计理念

商业酒吧街建筑结合西双版纳傣族传统建筑的民族形式,加以现代构成手法,屋顶造型高低错落,立面形式丰富多变,娴雅质朴,各见精巧,展现独具个性魅力的傣家风情和浪漫风格。一层沿湖面建筑墙体采用仿文化石以及米色仿石涂料,稳重大方;二层、三层采用仿木涂料,自然怡人;屋顶采用传统的傣式建筑的屋顶形式,高低错落。精致的建筑的细部,错落的空间设计,赋予酒吧街傣式建筑的文化气息。

The building in the commercial bar street combines the national style of the traditional architecture of the Dai Nationality in Xishuangbanna with modern construction techniques. With scattered roof modeling and versatile elevation styles, the buildings are elegant and exquisite, revealing the romance of the Dai Nationality with its unique charm. Artificial cultural stone and beige stone-like painting are adopted for the 1F wall of lakeside buildings, rendering a feeling of steady and magnificence; wood-like painting is adopted for the 2F and 3F, rendering a feeling of nature and comfort; the traditional scattered roof form of the Dai architecture is adopted for the roof. The exquisite architectural details and scattered space design render the bar street with a breath of the Dai Architecture culture.

03-04 万达西双版纳国际度假区酒吧街建筑外立面
05 万达西双版纳国际度假区酒吧街夜景

04

05

05 XISHUANGBANNA WANDA PLAZA
西双版纳万达广场

时间：2015 / 09 / 26　　**OPENED ON**：26th SEPTEMBER, 2015
地点：云南 / 景洪　　　　**LOCATION**：JINGHONG, YUNNAN PROVINCE
占地面积：3.38 公顷　　　**LAND AREA**：3.38 HECTARES
建筑面积：7.83 万平方米　**FLOOR AREA**：78,300M²

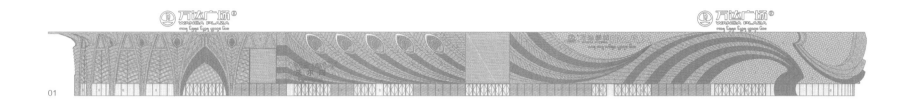

PROJECT OVERVIEW
广场概述

西双版纳万达广场作为万达首个文旅项目坐落于西双版纳州景洪市，占地约3.38公顷，总建筑面积7.83万平方米，地上3层及地下1层，是集购物、休闲、娱乐、亲子活动及餐饮等多种功能于一体的购物体验中心。

Xishuangbanna Wanda Plaza, the first cultural tourism project of Wanda Group, is situated in Jinghong, Xishuangbanna. Covering a site area of 3.38 hectares and a gross floor area of 78,300 square meters, the Plaza has three floors aboveground and one floor underground. It is an experience-focused shopping center that integrates shopping, leisure, recreation, family activities, food & beverage and other functions.

01 西双版纳万达广场立面图
02 西双版纳万达广场鸟瞰图
03 西双版纳万达广场总平面图

FACADE DESIGN
广场外装

设计灵感来自西双版纳最有魅力动物——孔雀。金色的铝板勾勒出孔雀的羽翼，蓝色彩釉玻璃呼应着天空，红色背景象征热烈而神秘的太阳。远观购物中心外立面，飞扬的弧线相互交融，形成流动的整体造型；弧形金色铝板和蓝色玻璃相互穿插，呈现出丰富的变化和动感；扭曲的受光面和背光面产生强烈的明暗对比与光影退晕变化，使立面呈现出丰富的变化和流线动态。

The design inspiration comes from the most attractive animal in Xishuangbanna - peacock. The golden aluminum plates depict the peacock's wing; the blue enameled glasses represent the blue sky; and the red background signifies the warm and mysterious sun. Looked from afar, the facade of the shopping center is full of dynamics, with interspersed arc-shaped golden aluminum plates and blue glasses. The warped surface with illumination and shade enriches the building façade with dramatic light-shadow contrast and gradient.

PART B | WANDA XISHUANGBANNA INTERNATIONAL RESORT 万达西双版纳国际度假区 | 061

06

04 西双版纳万达广场门头特写
05 西双版纳万达广场外立面
06 西双版纳万达广场入口

07

INTERIOR DESIGN
广场内装

室内步行街形体再现建筑的特色——中庭侧裙主造型高耸的人字形尖屋顶、多层叠加的屋檐,无不体现了当地的传统建筑特色,扶梯端头更是将叠加的特质发挥到极致;尖屋顶的柱状花式在细节之处又体现了佛教文化色彩。

The traditional architectural characteristics of local culture, have been integrated in the interior pedestrian street such as the gable pointed roof of the atrium side skirts and multilayer superposed eaves. The escalator ends displayed the superposition design characteristic even better. The pointed roof in columnar pattern represents the details of Buddhist culture.

08　　1F 时尚名品　　2F 潮流体验　　3F 休闲餐饮

07 西双版纳万达广场室内步行街
08 西双版纳万达广场商业落位图
09 西双版纳万达广场台阶
10 西双版纳万达广场景观雕塑
11 西双版纳万达广场景观

LANDSCAPE DESIGN
广场景观

景观设计灵感来源于西双版纳最具代表性的"泼水节"与"篝火节",也体现了"水"与"火"的演绎。广场铺装提取当地代表性的孔雀元素,引用孔雀羽毛的图案结合现代工艺材料,使得铺装既现代又具民族特色。通过优美的铺装曲线来到主广场,感受鼓灯造型的雕塑所奏响美妙的民族乐章。"铜鼓声声传山外,回音袅袅绕山涧",让人们感受到民族节日的气氛,不禁对生活产生美好的遐想。

The landscape design is inspired from "Water-Splashing Festival" and "Campfire Festival", the two most representative festivals in Xishuangbanna. The design has interpreted the concept of "water" and "fire". The plaza pavement employs typical local element-the peacock. The modern technology materials in peacock feather pattern contributes to a modern yet ethnic theme. Walking along the beautiful pavement toward the main plaza, one may hear appealing ethnic music played by the drum lamp-shaped sculptures. The landscape of Wanda Plaza brings the ethnic festival atmosphere and encourages the imagination of wonderful life.

C

WANDA PLAZAS
万达广场

WANDA COMMERCIAL
PLANNING 2015

01 TAIYUAN LONGFOR WANDA PLAZA
太原龙湖万达广场

时间：2015 / 09 / 30　　OPENED ON : 30th SEPTEMBER, 2015
地点：山西 / 太原　　　LOCATION : TAIYUAN, SHANXI PROVINCE
占地面积：44.22公顷　LAND AREA : 44.22 HECTARES
建筑面积：153.44万平方米　FLOOR AREA : 1,534,400M²

PROJECT OVERVIEW
广场概述

太原龙湖万达广场地处太原市杏花岭区，基地北部为北大街，西部隔龙潭公园临新建路，东部紧邻解放北路，南部至城坊街。广场占地面积44.22公顷，总建筑面积153.44万平方米，包括大型购物中心、超五星级酒店、超高层双塔写字楼、酒吧街和湖景豪宅等，是集购物、餐饮、文化、娱乐等多种功能于一体的大型商圈，也是活力四射的城市中心。

Located in Xinghualing District, Taiyuan City, Taiyuan LongFor Wanda Plaza faces North Street to the north, Xinjian Road to the west across Longtan Park, Chengfang Street to the south, and Jiefang North Road to the east. With a site area of 44.22 hectares and a gross floor area of 1,534,400 square meters, it includes a large shopping center, a super five-star hotel, a super-tall twin-tower office building, a bar street and lakeside mansions. The Plaza is a major retail precinct that integrates shopping, catering, culture, entertainment and other functions, resembling a vigorous city center.

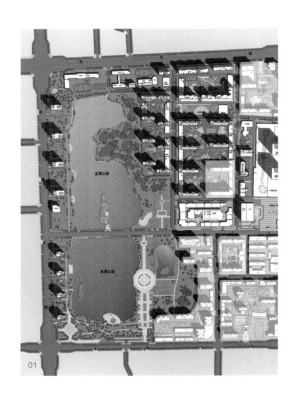

01 太原龙湖万达广场总平面图
02 太原龙湖万达广场建筑外立面

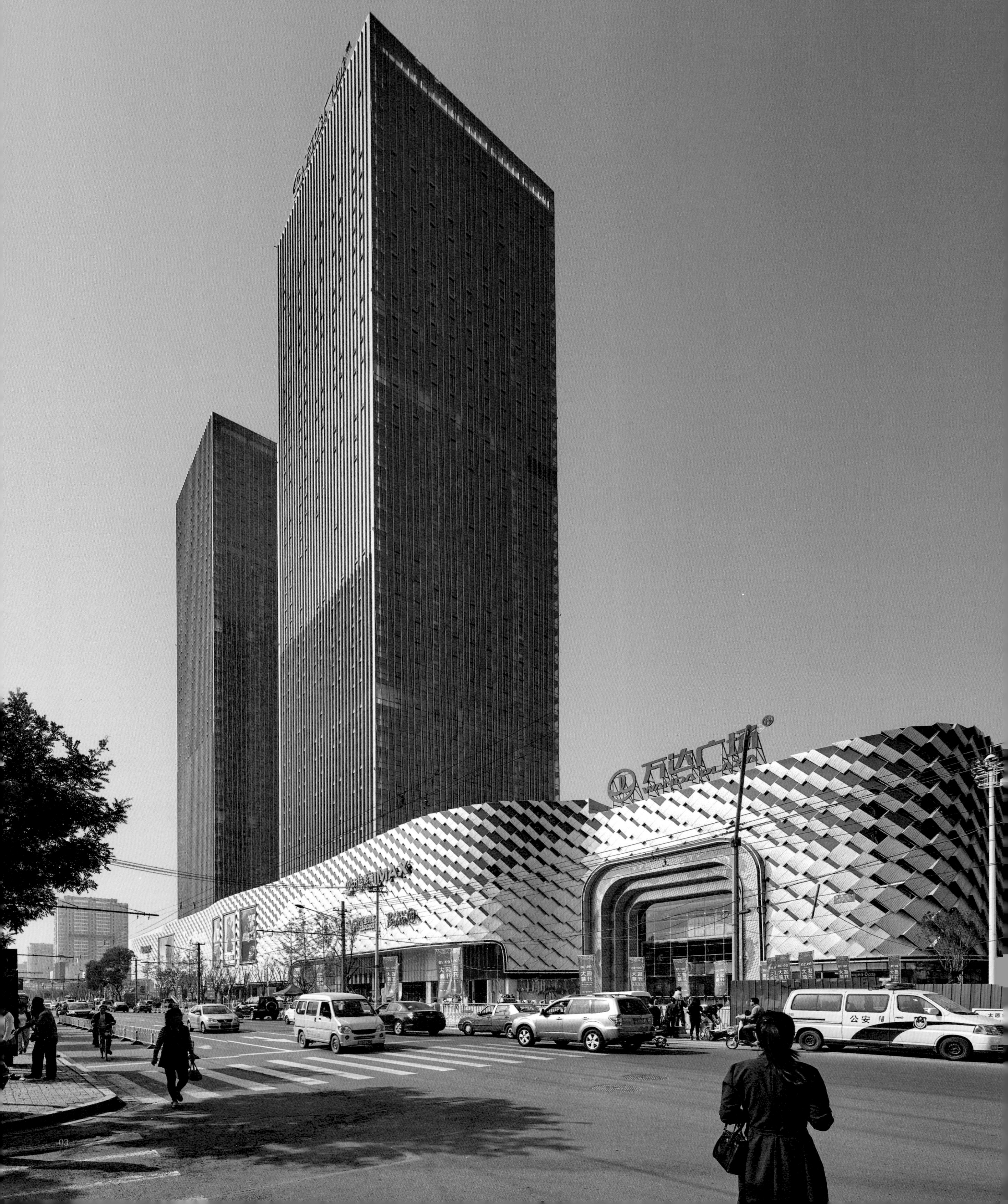

03

FACADE DESIGN
广场外装

太原为龙兴之城,古称"龙城",为体现龙城特色,建筑立面以拥有祥瑞之意的"龙"作为主题,抽象提炼设计的参数化"龙鳞"表皮,通过重复相搭接的45度方形鳞片,形成了宏伟的外立面设计,结合丰富的细节处理及富有"飞龙乘云"寓意的门头造型与"一店一色"的商业面,使得它正如一条巨龙"盘"在城市中央。塔楼上交织的线条,随着光影和视线的变化映衬出云彩图案,形成"龙盘云绕"的美景。

Taiyuan was the capital or provisional capital of many dynasties in China, thus it has the name of "Dragon City". To demonstrate its characteristics, the facade takes the auspicious "Dragon" as its theme. Through repeatedly lapping abstract parametric square dragon scale in 45 degree, a magnificent facade is built. With rich details, a gateway design implying "dragon rides on cloud reaching the sky" and "One Shop, One Style" shop-front design, the facade looks like a great dragon "wreathing" on the city center. The interweaving lines on the towers resemble a cloud pattern with changing lighting and sightlines, presenting a beautiful scene of "dragon wreathing on the cloud".

03 太原龙湖万达广场外立面
04 太原龙湖万达广场门头

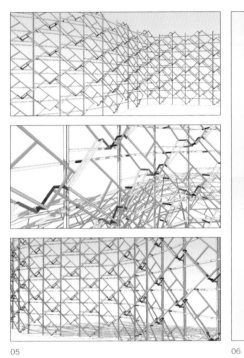
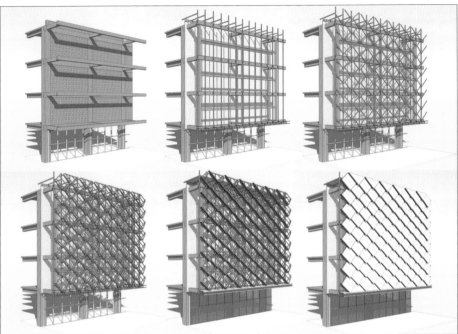

05 太原龙湖万达广场弧形龙鳞骨架
06 太原龙湖万达广场外立面结构
07 太原龙湖万达广场外立面特写
08 太原龙湖万达广场外立面特写
09 太原龙湖万达广场外立面

建筑外立面为完整的二维曲面，肌理是重复相互搭接的45度方形鳞片，通过参数化设计，每个方形鳞片的玻璃与金属比例不同，从而在强化二维曲面光影变化的同时形成抽象的"龙鳞"表皮。

The facade is an integrated 2D surface with a texture of repeatedly lapped square scales in 45 degree. Through parametric design, each square scale has a different glass-metal proportion, which enhances the 2D surface with changing lighting and forms an abstract dragon scale appearance.

10 太原龙湖万达广场商业橱窗
11 太原龙湖万达广场商业橱窗
12 太原龙湖万达广场商业橱窗立面图

太原龙湖万达广场创造性的采用了双层橱窗设计，同时确立了"一店一色"的设计处理理念。通过多方案的比较和实体模型分析，最终选取了四种橱窗的做法，排列变化形成完整丰富的立面商业表达，增加了近人尺度的设计细节，极大地丰富了建筑的沿街界面。

The Plaza creatively adopted a double-layered store front design and established a concept of "One Shop, One Style". After comparing several schemes and analyzing physical models, four types of store front were selected eventually. Through layout variation, it established an integrated and rich commercial expression, strengthened human scale design detail, and greatly enriched the building's interface along the street.

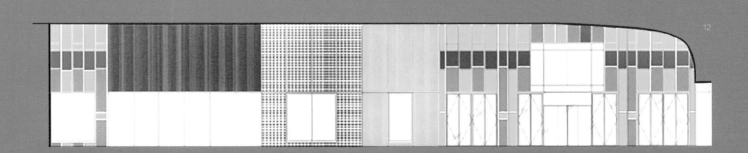

13

14

13 太原龙湖万达广场圆中庭
14 太原龙湖万达广场椭圆中庭
15 太原龙湖万达广场室内步行街顶棚
16 太原龙湖万达广场室内步行街顶棚

INTERIOR DESIGN
广场内装

室内设计将建筑外立面的语言引入室内,以"龙"为设计主线,形成室内外设计的"一体化"关系,具有连贯、统一、协调的特点。侧帮采用蚀刻金属板与玻璃结合,形成高低错落之势,意为"游龙戏水";在中庭顶部显示屏内做出了一个球体,取意"游龙戏珠";侧帮GRG、顶棚三角形的灯具、长街顶棚金属板均为龙鳞造型,形成了具有强烈艺术氛围的商业购物环境。

Interior design inherited facade design language and continues to take "dragon" as its theme. Thus an integrated exterior and interior design coherence, unification and coordination is achieved. The lateral wall combines etched metal plates with glass to form a staggered pose, meaning "dragon playing with water". A ball shaped display screen is placed at top of the atrium, meaning "dragon playing with a ball". GRG lateral wall, triangle-shaped ceiling lamps, metal ceiling plates of pedestrian street all employed dragon scale shape, contributing to a commercial shopping environment with artistic atmosphere.

15

16

PART **C**   WANDA PLAZAS
万达广场

19

4F   万达影城

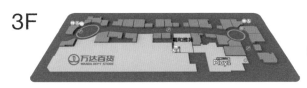
3F   万达百货
       家庭娱乐中心

2F   万达百货
       万达宝贝王
       US

1F   万达百货
       优衣库
       H&M

18

17 太原龙湖万达广场室内步行街
18 太原龙湖万达广场商业落位图
19 太原龙湖万达广场室内步行街

ONE STORE, ONE STYLE
一店一色

20-24 太原龙湖万达广场特色店面

LANDSCAPE DESIGN
广场景观

景观以"行云流水,如意龙城"为设计主题,景观设计致力打造成为富有品质、时尚、有情怀的城市综合体景观环境。景观不仅与建筑形式相互呼应、相互衬托,而且融入了当地浓郁的地域文化,将古建筑城隍庙与商业中心融为一体,历史感与现代感互相碰撞、相互融合;广场雕塑同地域文化结合,依据太原出土的北齐文物"红陶牛"实物形象,用黄铜铸成。利用现代景观的处理手法,打造出具有时代感又不乏人文底蕴的商业环境。

Following the theme of "Freely Flowing Cloud and Water, Wishful Dragon City", landscape design strives to create an urban scale landscape that is full of quality, fashion and romanticism. To this end, the landscape design not only echoed with the architectural style, but also added some local cultural elements. It combined the ancient old temple and the commercial center, resulting a collision and fusion of the modern and the historical. To reflect local culture, the Plaza sculptures are made of brass by referring to the physical image of "red pottery cattle", an antique of Northern Qi Dynasty unearthed in Taiyuan. Utilizing modern landscape techniques, a commercial environment featuring a sense of history and a sense of humanistic connotations is made available.

NIGHTSCAPE DESIGN
广场夜景

夜景照明准确把握建筑创意，采用柔和的灯光创造一个具有地域文化特征的、活跃的、富有朝气的现代商业体，赋予建筑"龙"的文化底蕴。绚丽的色彩变幻使龙鳞宛若晶莹剔透的琉璃一般美轮美奂、光彩夺目，呈现了立体丰富的照明效果，尽显"光艺术"与"高科技"融合的"超现实之美"，仿佛一场光与影的视觉盛宴。

Understanding the architectural concept, the nightscape design uses soft lighting to build a modern, vibrant and vigorous commercial building that is filled with regional and cultural characteristics. It gives the building the cultural deposit of "dragon". Gorgeous color change makes dragon scales look crystal clear, like colored glaze, magnificent and dazzling. Like a visual feast highlighting light and shadow, nightscape fully presents a brilliant 3D lighting effect of "surreal beauty" mixing "light art" and "high technology".

25 太原龙湖万达广场景观
26 太原龙湖万达广场花坛
27 太原龙湖万达广场花坛草图
28 太原龙湖万达广场夜景

02 CHONGQING BA'NAN WANDA PLAZA
重庆巴南万达广场

时间：2015 / 10 / 30 **OPENED ON**：30th OCTOBER, 2015
地点：重庆 / 巴南 **LOCATION**：BA'NAN DISTRICT, CHONGQING
占地面积：5.84 公顷 **LAND AREA**：5.84 HECTARES
建筑面积：24.98 万平方米 **FLOOR AREA**：249,800M²

01 重庆巴南万达广场鸟瞰图
02 重庆巴南万达广场总平面图
03 重庆巴南万达广场外立面

PART C WANDA PLAZAS 万达广场

PROJECT OVERVIEW
广场概述

重庆巴南万达广场位于重庆市巴南区龙洲湾街道地块内西北角，西靠渝南大道，东侧和南侧均为市政道路，轻轨三号线路经本地块，交通便利。广场由购物中心、室外商业街及甲级写字楼组成。总建筑面积超过99万平方米，是集购物、餐饮、娱乐、文化于一体的巴南商业新地标。

Chongqing Ba'nan Wanda Plaza is conveniently located in the northwest corner of Longzhouwan Subdistrict of Ba'nan District, a main district of Chongqing, facing Yu'nan Road to the west, municipal roads to the east and south, with Light Rail Line 3 passing by. The Plaza is composed of a shopping center, an outdoor commerical street and a grade A office building. With a gross floor area more than 990,000 square meters, it sets to be a new commercial landmark in Banan District that integrates shopping, catering, recreation and culture.

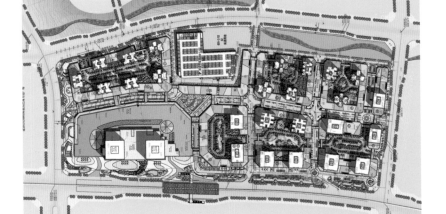

04

05

FACADE DESIGN
广场外装

重庆巴南万达广场的设计理念"抽象于山城连绵起伏的形态"、"融合了经典大气的几何图案"、"整合了穿插与组合的构成手法"。悬浮于整体框架之上的大面积的金色铝板幕墙,使得建筑极具动感。这条动感主线承前启后,串联起一块块的多样性表皮,而每块细节都各具特色,却又能完美地包容在一个整体中——刚烈、柔和、灵动——多种性格协调而统一。

The design concept of the project took the abstracted undulating form of the mountain city, infused with classic and grand geometric patterns, and integrated compositional methods of alternation and combination. The huge golden aluminum plate curtain wall that's floating above the entire frame builds a dynamic architectural volume. This dynamic theme is observed through connecting diverse skins in blocks, each containing distinct features and being perfectly melted into the whole facade, tough, soft and flexible. Different characters are thus coordinated and unified.

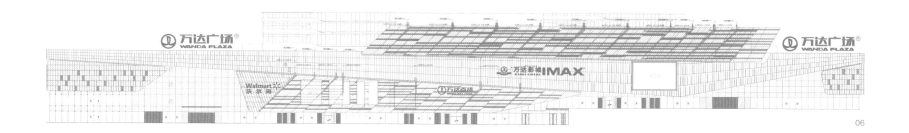

04 重庆巴南万达广场2号门头
05 重庆巴南万达广场外立面
06 重庆巴南万达广场立面图
07 重庆巴南万达广场1号门头

INTERIOR DESIGN
广场内装

3F

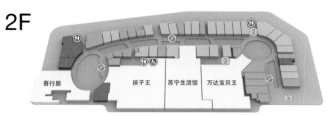

美时佳
海派健身
苏宁生活馆
大玩家

2F

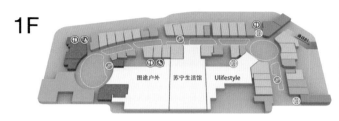

吾行居
孩子王
苏宁生活馆
万达宝贝王

1F

图途户外
苏宁生活馆
Ulifestyle

中庭深浅木色变换，搭配"巴南文化"元素的美陈与小品，在如"浪漫星空"的采光顶照射下，营造繁华的商业气氛；内街"天桥"犹如精心装饰的礼盒，更能提升购物空间的质感。入口门厅简洁大气；天花如明亮的"星河"，具有强烈的引导性；电梯厅和室内扶梯注重质感和细节，使人流在迅捷高效的行进间领略通透惊艳的商业氛围。

With deep and light wood color change, and the display of landscape articles containing "Ba'nan Culture", the atrium created a prosperous commerce atmosphere under the "romantic starry sky"-like day-lighting ceiling. The "overpass" at interior street looks like an elaborately decorated gift box, helping to enhance the quality of the shopping space. The entrance hallway is concise and grand. The brilliant "starry river"-like ceiling is quite directional. The elevator hall and interior escalator highlight quality and detail. All of these enable people to appreciate transparent and amazing commercial atmosphere while quickly passing by.

08 重庆巴南万达广场椭圆中庭
09 重庆巴南万达广场椭圆中庭
10 重庆巴南万达广场商业落位图

LANDSCAPE DESIGN
广场景观

景观设计元素统一并与建筑相得益彰——景观的精神是融合、体会并表现"场地的精神"。大体量的雕塑，远观时与建筑遥相呼应，走近时呈现出完全不同却非常震撼的效果。当人在景观中游走时，感知到的是其亲人尺度、肌理或者空间功能。合理利用地形条件并增加场地层次感——利用地形的高低变化，形成丰富的空间，在不同尺度之间相互转换。

Landscape elements are designed to be unified and harmonious with the building. The landscape design concept is the infusion, understanding and expression of "the spirit of site". The large sculptures echo with buildings afar while impress people when watching closer. Therefore, while wandering in the landscape, what we feel is its amicable scale, texture and spatial function. Through rationally utilizing the rolling topography and enhancing site grading, an abundant spatial inter-converting at diverse scales are witnessed.

NIGHTSCAPE DESIGN
广场夜景

照明设计构思围绕"山水之城"的主旨,突出巴南区唯美、清新、自然、生动的城市形象;灯光整体色彩绚丽,动态效果丰富,体现"巴山夜雨,碧水山青"的夜景照明设计主题。大商业立面使用具有力度的直线构成3个大面的体块。位于中间部位的是亮香槟色主体块,灯光表现出竖向线条,以"线"贯穿全场形成"面",形成视觉冲击力的夜间商业氛围;另外两个浅灰色体块,突出横向线条。这种"横向"与"竖向"对比强烈的处理,烘托了现代商业特有的速度感与节奏感。

Centering on the keynote of "landscape city", lighting design endeavors to mould an aesthetic, fresh, natural and lively city image. With radiant colors and dynamic effects, the nightscape lighting design took the theme of "night rain in Sichuan, blue water and green hill". The facade of large commercial is divided into three blocks with powerful straight lines. The main block in bright champagne, also the major area of lighting design, lies in the middle. In this block, lights are used to draw vertical lines that form a "surface", creating visually shocking commercial atmosphere at night. Whereas the other two blocks in light grey depict horizontal lines. Through such a "horizontal" and "vertical" contrast, the sense of speed and rhythm specific to modern commerce is highlighted.

OUTDOOR PEDESTRIAN STREET
室外步行街

重庆的夏天是炽热的,重庆的人是热情的。设计以重庆的热烈为基调,在室外步行街中充分运用各种色彩元素、多种材料,分段组合造型,营造热闹、丰富、绚丽的气氛;同时在一层商铺外增设外摆,增加商户经营价值的同时也大大提高步行街的人气,起到聚集人流的效果。

Chongqing is hot in the summer, Chongqing people are hospitable. So the design of Outdoor Pedestrian Street took hotness as its keynote. The design makes full use of diverse color elements and materials to apply segmented modeling, attempting to build lively, abundant and magnificent atmosphere. Meanwhile, stores on the ground floor added exterior furnishings along the street, to improve the popularity of the pedestrian street and effectively gather the flow of people.

11 重庆巴南万达广场景观雕塑
12 重庆巴南万达广场景观绿化
13 重庆巴南万达广场夜景
14 重庆巴南万达广场室外步行街

03 DALIAN KAIFAQU WANDA PLAZA
大连开发区万达广场

时间：2015 / 08 / 29　　OPENED ON : 29th AUGUST, 2015
地点：辽宁 / 大连　　　LOCATION : DALIAN, LIAONING PROVINCE
占地面积：9.87公顷　　LAND AREA : 9.87 HECTARES
建筑面积：73万平方米　FLOOR AREA : 730,000M²

PROJECT OVERVIEW
广场概述

大连开发区万达广场位于大连市金普新区，区位优势明显，占地9.87公顷，总建筑面积73万平方米。广场由购物中心、室外商业街、甲级写字楼、五星级酒店及公寓组成。其中购物中心建筑面积15万平方米，五星级酒店建筑面积3.68万平方米，写字楼建筑面积约12.5万平方米。

Situated at Jinpu New District, Dalian City, Dalian Kaifaqu Wanda Plaza enjoys supreme location advantage. The Plaza has a site area of 9.87 hectares and a gross floor area of 730,000 square meters, including 150,000 square meters of shopping center, and 36,800 square meters of five-star hotel 36,800 square meters of grade-A office building. It's composed of a shopping center, an outdoor commerical street, a grade-A office building and a five-star hotel.

01 大连开发区万达广场总平面图
02 大连开发区万达广场鸟瞰图

03

FACADE DESIGN
广场外装

广场设计构思契合"天使之翼,轻舞飞扬"的概念,通过起伏层叠的波浪形立面传递出立体构成的建筑美感,展现出滨海特色的动感。外立面幕墙通过系列的形体变化、不同层次的材质与色彩变化,给静态建筑立面赋予动感欢快的商业氛围,使建筑造型的商业感与现代感完美地融合。

Clinging to the idea of "Wings of Angle, Flying & Dancing", the Plaza design applied rolling wave-shaped facade to deliver 3D compositional beauty and unique dynamism of the coastal city. Benefiting from a series of shape variation and different gradations of materials and colors, the facade curtain wall gives the static facade a vigorous and cheerful commerce atmosphere. In such a way, the architectural retail sense and modern sense live in perfect harmony.

03 大连开发区万达广场外立面
04 大连开发区万达广场入口

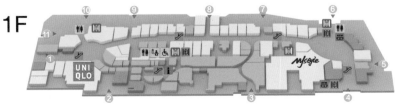
4F

3F

2F

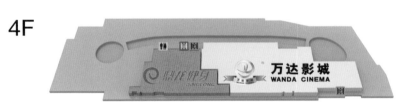
1F

服装　精品　餐饮　体验

INTERIOR DESIGN
广场内装

内装风格契合"海浪"的主题，充分体现了大连的地域特色，与建筑外立面相互呼应，取得内装、外装整体协调统一的效果；并且色彩和谐，线条优雅，创造出了大气、灵动的内装效果。内街连桥及扶梯的异型铝板、GRG等材料配合紧密，内藏灯带的运用恰到好处，既保证了整体的协调又丰富了设计细节。地面铺装体现出"海浪"的旋律，与天花、墙面造型中的"点、线、面"设计元素形成有机组合，形成了空间上的上下呼应，相得益彰。

The interior design adopted the theme of "wave" to fully demonstrate the regional features of Dalian. It echoed with facade design, in order to achieve an overall coordinated and unified effect. With harmonious colors and elegant lines, the design contributes to a grand and flexible decoration effect. The materials of the connected bridge and escalators over interior streets, such as abnormally-shaped aluminum plate, GRG and others, are in close coordination. Built-in lighting strip is properly arranged. Both elements guaranteed a harmonious overall effect and enriched design details. Floor pavement embodies the "wave" melody. It forms an organic combination with the design elements of ceiling and wall, "points, lines and surfaces", presenting a coherent and coordinated upper and lower space.

05 大连开发区万达广场圆中庭
06 大连开发区万达广场室内步行街电梯
07 大连开发区万达广场室内步行街连桥
08 大连开发区万达广场商业落位图

LANDSCAPE DESIGN
广场景观

景观设计以"金石海滩"为主题,着力打造"海风文化"的商业景观场所。大商业"带状广场"是由与建筑相呼应的波浪区域围合而成,象征金沙海滨的特色。雕塑诠释的是金色"海洋世界"的绚丽多姿,主雕通过蓝色海浪与金色海豚结合,表现充满吸引力的大商业标识物。

The landscape design centers on the theme of "Golden Stone Beach", and strives to build a retail landscape space featuring "Ocean Breeze Culture". The "Strip Plaza" of large commercial area is enclosed by a "wave" area that corresponds to buildings, embodying the characteristics of gold sand shore. The sculptures try to interpret a colorful golden "sea world". Blue sea wave and golden dolphin were put together in the main sculpture, to display the appealing icon of large commercial area.

NIGHTSCAPE DESIGN
广场夜景

夜间动感的灯光效果犹如"海浪"优美的波纹；入口的穿孔板以像素点表现，好像海面微波荡漾，泛起点点星光；红色陶板在LED小投光灯的照射下，光斑犹如盛开的美丽花瓣。通过静态表现的红色陶板与动态表现的金色铝板的流线穿插，取得时尚律动的效果，形成夜景照明的突出特色。

The dynamic lighting effect at night look like graceful ocean waves; perforated plates at the entrance were expressed in pixel dots, which look the waves in the rippling ocean, with blinking stars; the red ceramic plates, under LED projection lights, formed light spots that look like blooming flowers. Through the interweaving of static red ceramic plates and dynamic golden aluminum plates, a stylish and pulsing effect is achieved and the nightscape lighting feature is highlighted.

09 大连开发区万达广场景观花坛
10 大连开发区万达广场景观雕塑
11 大连开发区万达广场夜景
12 大连开发区万达广场室外步行街

OUTDOOR PEDESTRIAN STREET
室外步行街

"金街"以"金色海洋"为概念，通过铺装的拼砌，设计出灿烂的金色浪花纹路与活跃的海洋鱼类交相呼应。"金街"花池则诠释沉落于海底的历史瓷器瑰宝，展现金州人文的悠久文化。

Golden Street follows the idea of "Golden Ocean". With the cladding of pavements, brilliant golden waves echo with vibrant ocean fishes. Flower bed at the golden street represents the valuable ancient porcelains that were sunken at the bottom of the sea, attempting to showcase the long-standing culture of Jinzhou District.

04 SHANGHAI JINSHAN WANDA PLAZA
上海金山万达广场

时间：2015 / 07 / 17　　**OPENED ON**：17th JULY, 2015
地点：上海 / 金山　　　**LOCATION**：JINSHAN DISTRICT, SHANGHAI
占地面积：11.4 公顷　　**LAND AREA**：11.4 HECTARES
建筑面积：45.77 万平方米　**FLOOR AREA**：457,700M²

PROJECT OVERVIEW
广场概述

金山区地处杭州湾畔，位于沪、杭、甬及舟山群岛经济区域中心，是上海市的西南门户。丰富的土地资源、广阔的海岸线和建深水港的天然条件，构成了其得天独厚的地理优势、环境优势和经济辐射优势。广场紧邻金山区政府，周边的住宅区分布密集，使得该项目具有巨大的商业潜力。基地紧靠松卫南路和深海高速，交通便利。

Jinshan District is located at the Hangzhou Bay. It's the southwest portal of Shanghai and it's at the center of the economic zone of Shanghai, Hangzhou, Ningbo and Zhoushan Islands. With abundant land resources, vast coastline and favorable natural conditions for deepwater port, Jinshan District enjoys advantageous predominance in geography, environment and economic radiation. The project has great commercial potentials due to its adjacency to Jinshan District Government and the densely distributed residential areas around. It also enjoys convenient transportation as it borders Shenyang-Haikou Expressway and Songwei South Road.

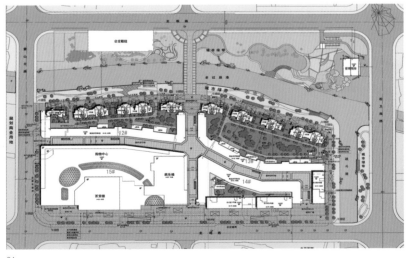

01 上海金山万达广场总平面图
02 上海金山万达广场外立面

03

FACADE DESIGN
广场外装

金山万达广场全过程精细化设计，无论是从整体的规划、景观绿化还是建筑材料，都反复修改调整，力求打造完美的万达广场。尤其购物中心外立面设计是以海浪为设计灵感，突出金山区滨海的独特地理优势。采用高档金属材质，创造出海天呼应、海水涟漪的错落层次感。

To build a perfect project, the Plaza applied fine design in the entire process ranging from overall planning, landscape planting to construction materials, which are all subject to repeated modification and adjustment. The facade design of shopping center is inspired by sea wave, trying to emphasize the unique geological advantage of coastal Jinshan District. Its application of high-grade metal created a staggered gradation of echoed sea-sky and rippling seawater.

FACADE DESIGN
广场内装

设计将古老港口的鲜明地域文化融入其中，表现"海上文明"和"海上丝绸之路"的起点，展望"新的海上丝绸之路"。步行街内装设计取材当地四通八达、活力无限、多姿多彩的个性，将"海舞风畅"定型为本案的主题线索，同时加入了"光辉照四方"延伸理念。无论是侧裙起伏的造型、玻璃的波浪图案，还是图案拼接、座椅、小品等细节，均提取"金山语言"，使得建筑之下涌动一股文化的力量。

Interior design incorporates distinct regional culture of the ancient harbor, to display the starting point of Maritime Silk Road and Maritime Civilization, and to envision the "New Maritime Silk Road". Sourcing its ideas from the all-round, energetic and colorful individuality of Jinshan District, the interior design of pedestrian street took "dancing sea and smooth wind" as the clue for its theme, and "shining far and wide" as its extention. "Jinshan Language" is found everywhere in the interior, from undulating modeling of side skirt, wavy pattern of glasses, to details such as pattern splicing, seats and landscape articles. Therefore a sense of culture is felt in the building.

03 上海金山万达广场1号入口
04 上海金山万达广场入口中庭
05 上海金山万达广场商业落位图
06 上海金山万达广场室内步行街

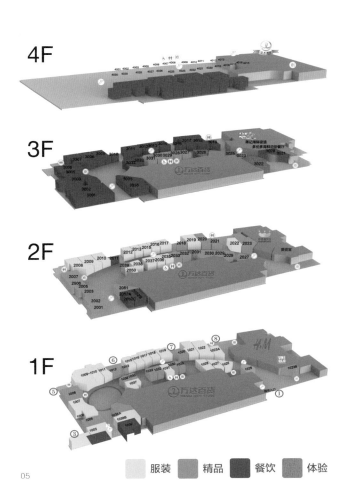

LANDSCAPE DESIGN
广场景观

景观设计追随建筑设计灵感，从丝绸入手，确定了"海上丝绸之路"的概念，然后提取浓缩为3个景观主题——"扬帆起航"、"乘风破浪"和"辉煌传奇"。每一个主题都有一个代表性的场景空间表达故事精神，同时在铺装、坐凳小品等细节处呼应建筑立面元素。景观空间以"场景化"原则，为参与和互动提供了条件。

Following the concept of architectural design, silk, the landscape design set "Maritime Silk Road" as its concept, and abstracted it into three themes, respectively being "Setting Sail", "Riding on the Wind" and "Glorious Legend". Each theme is provided with a representative scene space to deliver the spirit of the corresponding story. Besides, details such as pavement and benches echoed with facade design elements. Landscape space took "scenario" as the principle, to pave way for participation and interaction.

07

08

OUTDOOR PEDESTRIAN STREET
室外步行街

金山万达广场商业街富于创造性地采用了"海派建筑"的设计风格，将上海石库门风情与西方古典元素相结合，营造出特色鲜明的商业气息，吸引购物中心人流至此。商业街采用中国古典建筑常见的砖石元素，融合西方古典元素的斗拱柱式，加之玻璃和金属等现代元素，营造了"东方与西方文化兼容"、"古代与现代文明并蓄"的商业氛围。

The commercial street of the Plaza creatively employs "Shanghai-style Architecture". By combining Shikumen (a type of characteristic residential building in Shanghai) and classical western elements, the design created a featured commercial atmosphere to attract streams of shopping population. The commercial street adopted brick elements that are commonly seen in classical Chinese architecture, it also added some Western classical elements of column order and some modern elements of glass and metal, attempting to build a commercial atmosphere that incorporates Eastern and Western cultures and ancient and modern civilizations.

09

07 上海金山万达广场景观雕塑
08 上海金山万达广场水景
09 上海金山万达广场室外步行街雕塑小品
10 上海金山万达广场室外步行街

10

05 GUANGZHOU LUOGANG WANDA PLAZA
广州萝岗万达广场

时间：2015 / 07 / 17	**OPENED ON** : 17th JULY, 2015
地点：广州 / 黄埔	**LOCATION** : HUANGPU DISTRICT, GUANGZHOU
占地面积：8.98 公顷	**LAND AREA** : 8.98 HECTARES
建筑面积：41.41 万平方米	**FLOOR AREA** : 414,100M²

PART C　　WANDA PLAZAS
万达广场

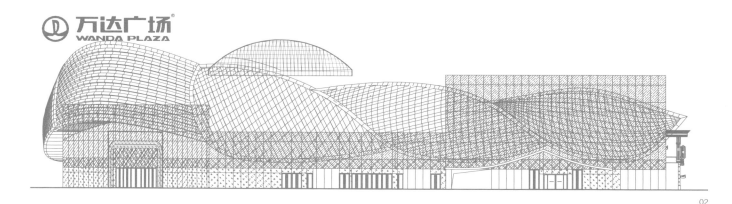

01 广州萝岗万达广场外立面
02 广州萝岗万达广场立面图
03 广州萝岗万达广场总平面图

PROJECT OVERVIEW
广场概述

广州萝岗万达广场位于广州市黄埔区，开创大道以南，科丰路以西，占地8.98公顷，总建筑面积41.41万平方米。广场由购物中心、室外商业街、甲级写字楼及公寓组成。其中购物中心建筑面积19.06万平方米，甲级写字楼建筑面积3.58万平方米，公寓建筑面积13.0万平方米，室外商业街5.60万平方米。

Guangzhou Luogang Wanda Plaza is located in Huangpu District, Guangzhou City. It's to the south of Kaichuang Avenue, and to the west of Kefeng Road. With a site area of 8.98 hectares, it has a gross floor area of 414,100 square meters, including 190,600 square meters of shopping center, 35,800 square meters of grade-A office building, 130,000 square meters of apartment, 56,000 square meters of outdoor commercial street.

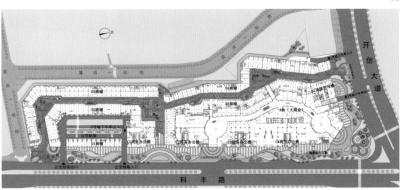

FACADE DESIGN
广场外装

广州市市花"木棉花"又名"红锦",树木高大,具有阳刚之美;花瓣形状饱满,以红色为主兼有橙色。广场外立面借鉴木棉花瓣的飘逸造型,打造流畅的双曲面造型。广州萝岗万达广场就这样精彩地绽放在广州人民面前。大商业立面用现代的设计手法,弧形线条和垂直线优雅相接,简约、时尚、轻盈,使人轻松愉悦。入口雨篷外挑,恰如空中划过的飘带,吸引人流进入商场,成为强烈的视觉中心。

Kapok (also called silk cotton), the city flower of Guangzhou, grows from a tall tree with masculine beauty. Its petals are mainly in red mixed with orange. Imitating the elegant kapok pedal shape, the facade design adopted a smooth double-curved surface modeling, to gorgeously bring the Plaza in front of Guangzhou people. The facade of large commercial area applied modern design method by elegantly connecting arc lines and vertical lines, to achieve concise, fashionable and light effects, which created a relaxing and pleasant atmosphere. The canopy at entrance looks like a ribbon flying across the sky, attracting pedestrians that are flowing into the stores. It becomes a strong visual center.

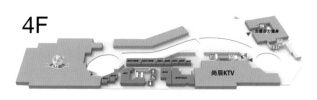

服装　精品　餐饮　体验

04 广州萝岗万达广场1号入口
05 广州萝岗万达广场2号入口
06 广州萝岗万达广场商业落位图
07 广州萝岗万达广场室内步行街连桥
08 广州萝岗万达广场椭圆中庭

INTERIOR DESIGN
广场内装

广州萝岗万达广场内装提取岭南建筑中"墙"的形态，并加以现代演绎。潇潇洒洒的木纹铝板造型与顶棚，在偌大的中庭空间上显得极为时尚与温馨。大胆地将侧裙与顶棚结合，营造出富有动感和节奏的环形动线效果。扶梯在空间中担当"主角"，极为动感的木纹铝板造型贯穿扶梯侧面，与地面铺装形成完美的效果。

The interior design of the Plaza tries to interpret the "wall" form found in "Lingnan Architecture" in a modern way. The stylish wood pattern aluminum plate and ceiling look remarkably chic and cozy in the spacious atrium. Through boldly combining side skirt with ceiling, a circular flow line full of dynamism and rhythm is presented. Escalator, the "main character" in the space, adopts the dynamic wood pattern aluminum plate on its sides to perfectly match the floor pavement.

LANDSCAPE DESIGN
广场景观

设计以"生命绽放"为主题,结合广州本地文化以及科学城的特殊地理位置,展现场地蓬勃的活力与向上发展的精神面貌。设计以流畅的曲线串联各个功能区域,表现花开绚烂时的魅力,以雕塑表现生命绽放的过程。明快的色调与建筑协调一致,使景观与建筑融为一体,契合了"花城广州"的灿烂、活力及生命的力量。

In order to show the thriving vitality and the uprising spiritual outlook of the site, the landscape design carried forward the theme of "Life Bloom" and took the local culture and special location of Guangzhou Science City into consideration. Smooth curves are used to connect each functional area to display the charm of blossom; sculptures are utilized to demonstrate the process of bloom; lively colors well integrated the landscape with the buildings. All of these comply with the magnificence, vigor and vitality of Guangzhou, the city of flowers.

09 广州萝岗万达广场景观雕塑
10 广州萝岗万达广场花坛喷泉
11 广州萝岗万达广场室外步行街
12 广州萝岗万达广场室外步行街雕塑小品

09

OUTDOOR PEDESTRIAN STREET
室外步行街

广州为多文化交汇地区,设计通过中西元素的结合、各种材质的混搭、造型各异的细节、丰富的美陈,营造热闹繁华的都市街道。

In response to the city's nature of culture interaction, the design incorporated Eastern and Western elements, diverse materials, differently-shaped details, and abundant display, so as to build a bustling city street.

06 DONGGUAN HOUJIE WANDA PLAZA
东莞厚街万达广场

时间：2015 / 11 / 06　　**OPENED ON** : 6th NOVEMBER, 2015
地点：广东 / 东莞　　　　**LOCATION** : DONGGUAN, GUANGDONG PROVINCE
占地面积：13.98 公顷　　**LAND AREA** : 13.98 HECTARES
建筑面积：61.48 万平方米　**FLOOR AREA** : 614,800M²

01 东莞厚街万达广场外立面
02 东莞厚街万达广场总平面图

PROJECT OVERVIEW
广场概述

东莞厚街万达广场位于东莞市厚街镇，处东莞版图中部，占地13.98公顷，建筑面积61.48万平方米。广场是集大型购物中心、商业休闲步行街、甲级写字楼、高端住宅及精装SOHO等于一体的城市综合体，包含万达影城、宝贝王、大玩家、电器卖场、大型超市等门类，以成熟的运营模式，携手国际一线品牌，构筑独立的大型商圈，缔造厚街商业新地标。

Dongguan Houjie Wanda Plaza is located in Houjie Town, the center of Dongguan City. With a site area of 13.98 hectares and a gross floor area of 614,800 square meters, the Plaza is a city complex that integrates large-scale shopping center, commercial pedestrian street, well-decorated SOHO, grade-A office building and high grade residence into one. It includes Wanda Cinema, Kidsplace, Super Player, electronic market and large supermarket. With mature operation model and first-line international brands, the project aims to build an independent large commercial area and to set a new commercial landmark in Houjie.

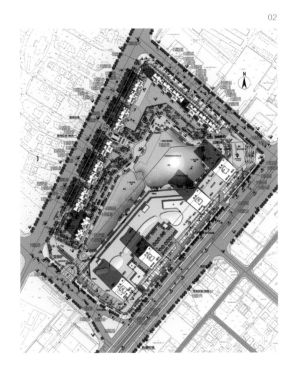

FACADE DESIGN
广场外装

运用现代的材料、简洁流畅的形体，生动地勾勒出"鱼跃于水"的生动画面，反映出广东人对水的理解和热爱。立面形体流畅，以沉稳的石材为底，上部搭配多变的铝板墙面，同时点缀灵动的玻璃材质，呼应了"大中见小"、"小中见细"、"细中见精"的建筑设计手法，取得了简洁而不简单的立面效果。大商业外立面采用了石材、玻璃和铝板三种主要材料——玻璃清透如水、石材质感如沙、铝板流线似鱼——展现了一幅"虚实结合"、"静中喻动"的画卷。

Thanks to modern materials and concise while smooth shape of the building, a picture of "Fish Diving in the Water" is vividly depicted, reflecting Cantonese's understanding and passion of towards water. The facade employed the design method of "small in big", "detail in small" and "refinement in detail" - sturdy stone at the bottom, changing aluminum plate wall at the top, flexible glass in the middle, resulting in a concise yet not simple effect. The facade design of large commercial area mainly adopts three materials, water-like-transparent glass, sand-textured stone and fish-streamlined aluminum plate, unfolding a picture of "virtual-real combination" and "dynamic-static balance".

03 东莞厚街万达广场外立面
04 东莞厚街万达广场1号入口

INTERIOR DESIGN
广场内装

广场在材质、形式上都运用了"钻石、切割"的概念。材质上,选了反光度高的材质;造型上,用了切割面的并接。这样处理,如同大小相间的钻石分布在空间的各个区域,营造了高档的购物氛围。商场的电梯底部采用三角形砂面不锈钢材质,如同钻石表面的质感。过桥采用蓝色不锈钢与铝板,两种材料有规律的拼接形成条纹造型,在空间中起了锦上添花的作用。侧裙由白色石膏板和条形的LED藏光构成,将整个空间连贯起来,凸显空间的结构之美。

The concept of "Diamond and Cutting" can be found both in materials (materials with high reflectance) and modeling (cutting surface slicing). In such a way, the whole space is distributed with diamonds of different sizes, contributing to a high-end shopping atmosphere. The bottom of the elevator in department store is made of triangle frosted stainless steel to simulate the surface texture of a diamond. Bridges regularly slice blue stainless steel and aluminum plates to form striped pattern, which further beautifies the space. Side skirt is composed of white gypsum board and hidden LED strip to bring the whole space together, highlighting structural beauty.

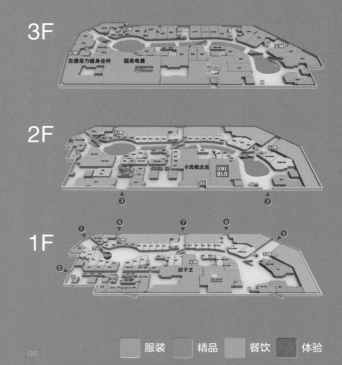

05 东莞厚街万达广场椭圆中庭
06 东莞厚街万达广场商业落位图
07 东莞厚街万达广场圆中庭

LANDSCAPE DESIGN
广场景观

景观以岭南地区的"客家文化"为主题，打造传承地域特色文化的商业景观场所。大商业带状广场由系列造型花坛、水景与大商业建筑围合而成。花坛造型优美，水景错落有致，结合南方特有的植物，打造出具有岭南地域特色的植物群落。金街内人工湖蜿蜒曲折，湖水清澈见底，与湖中心的拱桥相互衬托。景观雕塑将古代的"八景"加以提炼，与铁艺结合，表现原始的张力。牌坊取材于岭南民居及镬耳山墙，体现出了古代民居的魅力。

The landscape design took "Hakka Culture" in Lingnan area as its theme, attempting to build a commercial landscape space that inherits distinct local culture. Linear plaza in large commercial area is enclosed by a series of parterres, waterscape and large commercial buildings. The appealingly-shaped parterres and well-proportioned waterscape, together with endemic plants in the south, help to build a plant community with local characteristics of Lingnan. The meandering and crystal clear artificial lake in golden street and arch bridge in the center correspond to each other. Landscape sculptures extract ancient "Eight Sights" and use iron material to show the original tension. Memorial archway originated from folk house in Lingnan and pot handle-shaped gable unveil the charm of ancient folk house.

08 东莞厚街万达广场景观花坛
09 东莞厚街万达广场景观牌坊
10 东莞厚街万达广场室外步行街沿湖景观
11 东莞厚街万达广场室外步行街

OUTDOOR PEDESTRIAN STREET
室外步行街

多种风格的结合、多种形体的变化、多种材料的交接，交织形成了一条真正的商业街——它既有现代商业的性格，也有商业街独有的厚重感；极力表现出本土商业"生长"的状态，同时散发着强有力的生命。每一个单元能够自成一体、代表着一种文化，都可以作为一个独立的建筑而存在。

The combination of diverse styles, the change of different shapes and the connection of various materials contribute to a real commercial street. The street bears both character of modern retail and the sense of heaviness in commercial street; It strives to present a "growing" sense of local business and shows its strong vitality. Each unit stands by itself and represents a kind of culture, so as each unit can be served as an independent building.

07 LIUZHOU CHENGZHONG WANDA PLAZA
柳州城中万达广场

时间：2015 / 11 / 27　　**OPENED ON**: 27th NOVEMBER, 2015
地点：广西 / 柳州　　**LOCATION**: LIUZHOU, GUANGXI ZHUANG AUTONOMOUS REGION
占地面积：13.1 公顷　　**LAND AREA**: 13.1 HECTARES
建筑面积：66.80 万平方米　　**FLOOR AREA**: 668,000M²

PROJECT OVERVIEW
广场概述

柳州城中万达广场位于柳州市城中区文兴路与东环大道交叉口，占地13.1公顷，总建筑面积约66.80万平方米；其中地上部分51.50万平方米，地下部分15.30万平方米，包含商业购物中心、甲级写字楼、嘉华酒店、SOHO公寓、高级住宅及金街等业态。

Liuzhou Chengzhong Wanda Plaza is located at the intersection between Wenxing Road and East Ring Road, Chengzhong District, Liuzhou. The Plaza covers a site area of 13.1 hectares and a total gross floor area of 668,000 square meters, including 515,000 square meters of aboveground area and 153,000 square meters of underground area. It includes shopping center, grade-A office building, Wanda Realm Hotel, SOHO apartment, high grade residence, golden street and other commercial programs.

01 柳州城中万达广场总平面图
02 柳州城中万达广场鸟瞰图

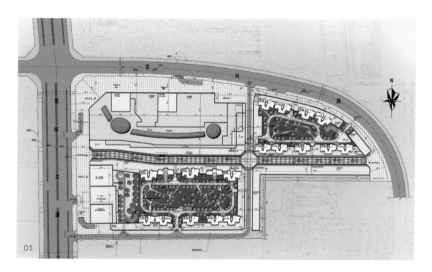

03

04

FACADE DESIGN
广场外装

龙在中国具有独特的文化地位，而柳州市也因"八龙见于江中"的典故被称为"龙城"。设计紧密结合这一特点，致力于打造具有"龙的形态"和"龙的精神"的城市广场——整个购物中心形态流动舒展，东高西低，以一种"龙抬首"的昂扬姿态挺立于城市道路一侧，极大地提升了城市形象。东、西两个主要出入口位于龙首和龙尾，重点打造的"IMAX 影院"位于龙首的核心部位。层层展开的弧线，给整个造型赋予腾飞之感；局部采用玻璃，表面覆盖"龙鳞"状肌理，使立面质感更加丰富，"龙"的整体形象更加丰满。

The dragon has a unique cultural status in China. Liuzhou city is named the Dragon City (it is said in folklores that eight dragons appeared in the Liujiang River). In response to this feature, the design is dedicated to building a city plaza with dragon shape and dragon spirit. The whole shopping center is erected on one side of the city road with a dragon head raising pose, smoothly stretching higher in east and lower in west, greatly promoting the city image. Two main entrances/exits on the east and the west stand at the dragon's head and tail, highlighting the core position of IMAX Cinema which lies in the dragon's head. Arc lines unfold layer by layer, presenting an image soaring high. Randomly applied glasses in "dragon scale" texture enriched the facade texture and better shaped the "dragon" image.

INTERIOR DESIGN
广场内装

设计遵循"呼唤自然,亲近生态"的理念,从宜居山水中寻找设计元素,突出融入式的艺术氛围;以"青山环绕,水抱城流"为母题,突出空间的纯粹与灵动;以融合和对比为基本概念对商业空间进行精心刻画。

Following the idea of "calling nature and approaching ecology", the interior design sources it's design elements from livable landscape and tries to build an immersive artistic atmosphere. Taking the theme of "mountains and waters around", the design attempts to present a pure and flexible space. Holding the basic concept of integration and contrast, the design carefully depicts the commercial space.

03 柳州城中万达广场1号入口
04 柳州城中万达广场外立面
05 柳州城中万达广场圆中庭
06 柳州城中万达广场商业落位图
07 柳州城中万达广场室内步行街

05

服装　精品　餐饮　体验

06

07

LANDSCAPE DESIGN
广场景观

景观围绕柳州"山奇·水秀·民族美"来描绘"越绝孤城千万峰，江流曲似九回肠"的"龙城柳州"；挖掘"青山环，绿水绕，多民族"的特色，打造荟萃自然景观与人文景观的独特广场环境。

Landscape design focuses on "mysterious hills, clear waters and beautiful ethnic" to portray the picturesque "Dragon City". The design further digs out characteristics of "green hills around, green waters around, multiple ethnics", to build a featured plaza environment that gathers natural landscape and cultural landscape together.

OUTDOOR PEDESTRIAN STREET
室外步行街

步行街立面设计综合考虑了商业活力的需求以及各店铺平面的相似性，决定遵循"一店一色"的设计理念，以混搭、拼贴等处理手法突出各个店铺自身的特点，从而塑造活跃的商业氛围。从民国、法式、ArtDeco等多种艺术风格中汲取灵感，设计了四十余种不同的单元类型，并精心协调其位置、尺度，形成了各具特色又完整统一的立面效果。

Given the demand for commercial vitality and the similarity in stores' layouts, the facade design of Pedestrian Street took the concept of "One Shop, One Style". To this end, mixing and splicing methods are employed to highlight respective features of each store, contributing to a vibrant commercial atmosphere. Besides, the design sources its ideas from diverse artistic styles, such as the Republican Period style, French style and Art Deco style, and more than 40 design elements are created. Through meticulously arranging and positioning these elements, a distinctive yet integrated facade is found.

08 柳州城中万达广场景观小品
09 柳州城中万达广场主雕塑
10 柳州城中万达广场室外步行街入口
11 柳州城中万达广场室外步行街地铺
12 柳州城中万达广场室外步行街

08 GUILIN GAOXIN WANDA PLAZA
桂林高新万达广场

时间：2015 / 09 / 12	**OPENED ON**: 12th SEPTEMBER, 2015
地点：广西 / 桂林	**LOCATION**: GUILIN, GUANGXI ZHUANG AUTONOMOUS REGION
占地面积：7.3公顷	**LAND AREA**: 7.3 HECTARES
建筑面积：33.0万平方米	**FLOOR AREA**: 330,000M²

PROJECT OVERVIEW
广场概述

桂林高新万达广场处于桂林市七星区穿山东路与环城南一路交汇处，交通便利。广场占地7.3公顷，总建筑面积33.0万平方米，由购物中心、室外步行街及住宅组成。其中购物中心建筑面积15.1万平方米，室外商业街建筑面积4.7万平方米，住宅及公寓建筑面积共约13.2万平方米。

Ideally located at the intersection between Chuanshandong Road and Huancheng South 1st Road in Qixing District, Guilin City, Guilin Gaoxin Wanda Plaza enjoys convenient transportation. The Plaza covers a site area of 7.3 hectares and a gross floor area of 330,000 square meters, including 151,000 square meters of shopping center, 47,000 square meters of outdoor commercial street and 132,000 square meters of residence and apartment.

01 桂林高新万达广场总平面图
02 桂林高新万达广场鸟瞰图
03 桂林高新万达广场立面图

PART **C**   WANDA PLAZAS
万达广场

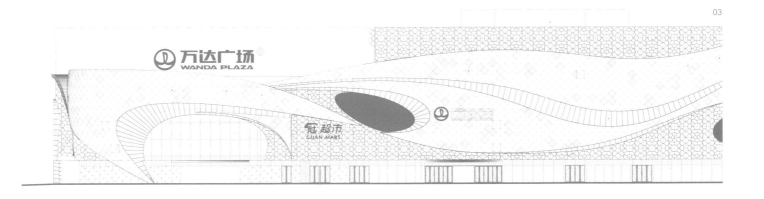

FACADE DESIGN
广场外装

以"山水云间"为题,演绎桂林"山行云流水"的神韵,运用现代设计手法传达出悠远的人文诗情。在造型上,大胆运用自由曲线,形成宛若河流奔涌或云出山林的飘逸姿态。晶莹剔透的彩釉玻璃幕墙,就像漓江流水,与群山、白云形成强烈的虚实呼应,将"山水云间"的意境建推向了顶峰。

The facade design took "Landscape in Clouds" as its theme in order to interpret the charm of "Drifting Clouds and Flowing Water" in Guilin. It employs modern design methods to deliver profound humanistic charm. It boldly uses free curves to shape the facade, forming an elegant picture of surging rivers or rising clouds from mountains. It uses crystal clear enameled glass curtain wall that looks like the running water of Lijiang River. The glass curtain wall becomes a strong virtual-real contrast with mountains and clouds, thus pushing the theme of "Landscape in Clouds" to the zenith.

04

04 桂林高新万达广场2号入口
05 桂林高新万达广场外立面
06 桂林高新万达广场椭圆中庭
07 桂林高新万达广场商业落位图

05

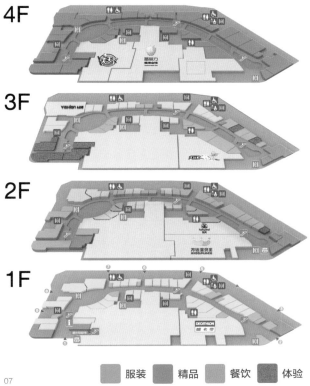

INTERIOR DESIGN
广场内装

内装设计营造"清影漓江"的意境，如乘轻舟游漓江，体验宁静、悠闲的购物环境。内街中的所有"连桥"都以"桥"的形式体现，保证内街的流畅性，配合地面蜿蜒的曲线造型，体现蜿蜒流动的漓江水；两侧侧板上的冲孔发光板，则体现"月色余辉"的美景。

The interior design tries to create an artistic conception of "Ligiang River's Light & Shadow", through which people may experience a quiet and leisure shopping environment as if they are touring Lijiang River by boat. The interior street adopts the form of real bridge as its connecting bridges to make it smooth; combined with wandering curve-shaped ground, it exhibits a flowing Lijiang River; the perforated luminous panel on both sides depicts a beautiful scene of "moonlight afterglow".

服装　精品　餐饮　体验

LANDSCAPE DESIGN
广场景观

景观以美丽的漓江及壮族传统文化为概念设计来源，着力打造一个传承地域特色文化的商业景观场所。大商业带状广场景观，融入波光粼粼的建筑立面元素，形成视觉上的延伸。雕塑设计以"山水·山音"为主题线索，通过系列人物的情景雕塑，展现壮族人民"漓江捕鱼"、"梯田耕作"、"山涧对歌"等生活场景，富于桂林地区民风特色。主景雕塑《山音律动》，宛如悦动的音符在山间传播开来，诠释桂林独特而优美的山水乡音和景色。

The conceptual design of landscape is originated from the beautiful Lijiang River and the traditional culture of the Zhuang Nationality, attempting to build a commercial landscape space that inherits features of the regional culture. Incorporating facade elements such as sparkling river, the linear plaza of large commercial area is visually extended. The sculpture design took "water from the mountains, sound from the mountains" as its thematic clue. The scenario sculptures of a series of people showcased daily life scenes of the Zhuang nationality, such as "fishing in Lijiang", "farming on terraced field", "singing in the mountains", which has ethnic features of Guilin area. The main sculpture "Sound and Rhythm of the Mountain" is like that the lively music notes are spreading among the mountains, to interpret the unique beauty of Guilin, mountains and streams, folk music and beautiful scenes.

OUTDOOR PEDESTRIAN STREET
室外步行街

步行街立面变化丰富,能够让消费者真正体会到"逛"的乐趣;充分挖掘桂林当地独有的地域特色,从"桂北民居"及"桂林山水"中吸取设计元素,经过艺术提炼与整合,使金街具有强烈的商业氛围和地域属性。

Consumers may truly experience the joy of "shopping" with the changeable facade of the pedestrian street. Fully utilizing on unique regional characteristics, drawing design elements from "Folk House in Northern Guangxi" and "Guilin Scenery", through artistic extraction and integration, the golden street is endowed with a strong commercial atmosphere and regional attribute.

08 桂林高新万达广场地铺
09 桂林高新万达广场主雕塑
10-12 桂林高新万达广场室外步行街景观小品
13 桂林高新万达广场室外步行街

09 GUANGZHOU NANSHA WANDA PLAZA
广州南沙万达广场

时间：2015 / 12 / 22　　**OPENED ON**：22ⁿᵈ DECEMBER, 2015
地点：广东 / 广州　　　　**LOCATION**：GUANGZHOU, GUANGDONG PROVINCE
占地面积：7.14公顷　　　**LAND AREA**：7.14 HECTARES
建筑面积：40.85万平方米　**FLOOR AREA**：408,500M²

PROJECT OVERVIEW
广场概述

广州南沙万达广场位于广州市南沙区环市大道与双山大道交界处西北侧，占地面积7.14公顷。地上建筑主体分为南区大商业（购物中心）7.89万平方米，北区外商铺（裙楼）5.0万平方米和6栋26~29层商务办公塔楼18.76万平方米，其他配套0.2万平方米；地下建筑面积9.0万平方米；总建筑面积40.85万平方米。

With a site area of 7.14 hectares, Guangzhou Nansha Wanda Plaza is located in Nansha District, Guangzhou. It's at the northwest corner of the intersection between Huanshi Avenue and Shuangshan Avenue. Its gross floor area reaches 408,500 square meters, including aboveground buildings and 90,000 square meters of underground programs. Aboveground buildings cover 78,900 square meters of large commercial area (shopping center) in south area, 50,000 square meters of stores (podium), 187,600 square meters of six 26~29-floor commercial office towers in north area, and 2,000 square meters of other ancillary facilities.

01 广州南沙万达广场总平面图
02 广州南沙万达广场外立面

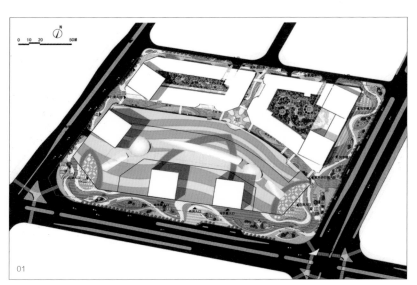

FACADE DESIGN
广场外装

广场立面取材"花城广州"的概念,用"花"的寓意反映城市繁荣、欣欣向荣的面貌。彩色铝板层层包裹的形态和错落交织的动态完美地再现了花蕾含苞待放的美态;立面流畅的曲线将"花丛舞动,万蕊浮花"表现得淋漓尽致。入口处流畅的曲线与立面表皮共生出挑的雨篷具有强烈的视觉冲击力。

Sourcing ideas from "Guangzhou, Flower City", the facade uses flowers to imply prosperous and flourishing city appearance. Staggered colorful aluminum plates wrap around the building to represent the beauty of budding flowers. The smooth facade streamline vividly depicts the scene of "flower dance & floating flowers". The flowing curve at the entrance and the overhung canopy extended from the facade skin generated a powerful visual shock.

03

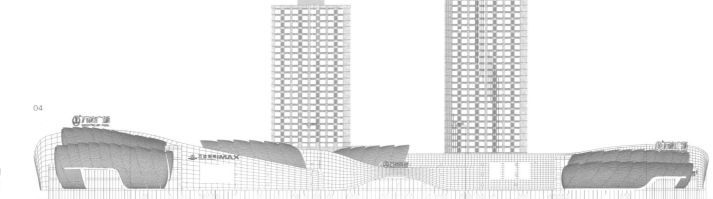

04

03 广州南沙万达广场2号入口
04 广州南沙万达广场立面图
05 广州南沙万达广场1号入口
06 广州南沙万达广场商业落位图
07 广州南沙万达广场室内步行街
08 广州南沙万达广场圆中庭

05

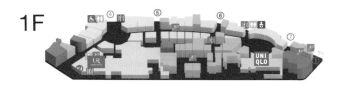

06

■ 服装 ■ 精品 ■ 餐饮 ■ 体验

INTERIOR DESIGN
广场内装

内装设计从建筑设计中提炼出横跨整个室内的自由曲线，犹如海浪般自由舒展、流动延伸。体现"水文化"的线条，在空间内形成交错的序列，从横向空间到纵向空间，线条均流动其中，成为空间构成的要素。

The interior design draws free curves from architectural design. They span the entire interior space, freely stretching and flowing like sea waves. The staggered lines exhibited "water culture", running through both transverse space and vertical space, becoming compositional spatial elements.

07

08

09

10

LANDSCAPE DESIGN
广场景观

景观在平面上采用"水墨莲花"的意向进行设计,加以虚实变化、"笔断意连"的表现,使整个平面联系紧密。金街的主题从"莲"延伸到"藕",以"藕的主题"概念呼应"莲的主题"概念,由此形成各有特点的表现空间,但又如同藕丝般紧密地联系在一起。在外街的主题雕塑采用"莲"为主题,在内街的雕塑小品采用"藕"为主题,以此在区分中形成变化。

Following the design idea of "Ink Lotus", the landscape incorporated virtual-real variation and theme-focused expression to render a closely linked plan. The golden street adopted the theme of "lotus root" to echo with the theme of "lotus". Accordingly it developed distinctive yet connected spaces, like lotus roots. The theme sculptures of outdoor street and interior street respectively took the "lotus" theme and "lotus root" theme for the purpose of variation.

OUTDOOR PEDESTRIAN STREET
室外步行街

室外金街在细部处理方面，通过仿石材西洋古典风格壁柱、拱券、檐口和线脚的运用，以及外墙面砖、仿木构架和遮阳格栅扇的穿插，体现出广州地方建筑特色，给人以亲切感。此外，醒目的彩色涂料、金属格栅、玻璃栏板、玻璃幕墙以及金属面板又使建筑平添了时尚气息。立面材质突出虚实、质感和色彩的对比，强化了商业氛围。

When designing the details, outdoor golden street applied both Western classical style, such as stone-like pilasters, arch, eave, molding, and local architecture features, such as exterior wall brick, wood-like frame and sun-shading grille, bringing a cordial feeling. Moreover, eye-catching color coated metal grilles, glass balusters, glass curtain wall and metal panels make the buildings fashionable. The facade materials employed contrast of virtual-real, texture, color, to intensify the commercial atmosphere.

11

12

09 广州南沙万达广场景观
10 广州南沙万达广场主雕塑
11 广州南沙万达广场室外步行街入口
12 广州南沙万达广场室外步行街
13 广州南沙万达广场室外步行街雕塑

10 NANNING ANJI WANDA PLAZA
南宁安吉万达广场

时间：2015 / 12 / 05　　**OPENED ON :** 5ᵗʰ DECEMBER, 2015
地点：广西 / 南宁　　**LOCATION :** NANNING, GUANGXI ZHUANG AUTONOMOUS REGION
占地面积：8.4 公顷　　**LAND AREA :** 8.4 HECTARES
建筑面积：58.2 万平方米　　**FLOOR AREA :** 582,000M²

01 南宁安吉万达广场总平面图
02 南宁安吉万达广场外立面草图
03 南宁安吉万达广场鸟瞰图

PROJECT OVERVIEW
广场概述

南宁安吉万达广场位于南宁市西乡塘区，北邻高新大道，由购物中心、室外商业街、写字楼、公寓及住宅组成。其中购物中心位于东地块，综合大商业五层（含室内步行街、购物、娱乐、餐饮等）；两幢乙级写字楼分别为26层和27层；另有临街商铺及配套服务设施。

Anji Wanda Plaza is located in Xixiangtang District, Nanning City. It's adjacent to Gaoxin Avenue in the north. The Plaza is composed of a shopping center, an outdoor commercial street, office buildings, apartments and residence. The shopping center is in the east; the large commercial area has five floors (interior pedestrian street, shopping, recreation and food & beverage included); the grade-B office buildings include one 26-floor and one 27-floor; the project also includes storefronts and supporting facilities.

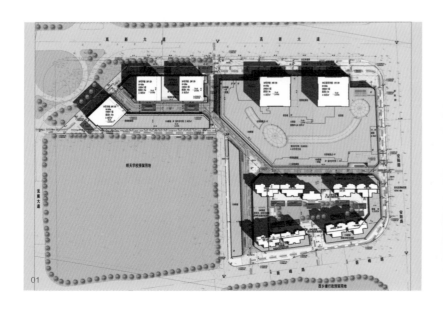

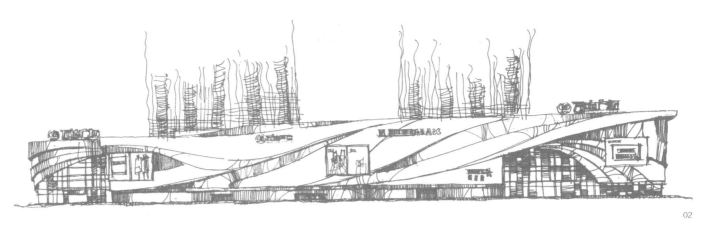

FACADE DESIGN
广场外装

建筑形体用曲线作为设计主要语言，连续流畅的白色铝板与金色穿孔铝板交相辉映，犹如河流和山川所展示出的自然姿态，大气磅礴；也如同自然界婀娜的绿色植物，灵动柔美。而这正是契合了南宁城市的面貌——奔流不息、欣欣向荣。

The architectural form is mainly delivered through curves. Continuous and smooth white aluminum plates and golden perforated aluminum plates interact with each other, just like the natural gestures that the rivers and mountains show, grand and magnificent; also like the graceful green plants in nature, flexible and soft. Such a design exactly echoes with the urban appearance of Nanning City-progressing and flourishing.

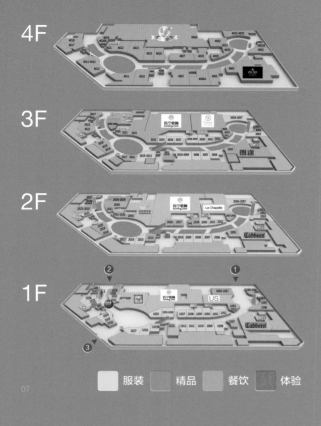

04 南宁安吉万达广场1号入口
05 南宁安吉万达广场外立面草图
06 南宁安吉万达广场外立面
07 南宁安吉万达广场铺位图
08 南宁安吉万达广场室内步行街连桥
09 南宁安吉万达广场圆中庭

INTERIOR DESIGN
广场内装

内装设计提取朱槿花标志造型并以钻石切割形体作为概念源泉。室内各区域的钻石切割造型分别以两种造型及两种颜色的材料构成,并通过混搭在简练中体现丰富的视觉效果。入口设计是内装设计的重中之重,为此运用大商业主体元素钻石切割形体造型,使入口区起到连接室内外的纽带作用。顶棚运用折面与斜线的混搭,产生了钻石状的造型,组成虚实相间的凹凸体块,使入口更有品质感,提升全场档次。

The interior design was inspired by Jaba and it took the diamond-cutting form as the origin for its concept. The diamond-cutting form on the interior adopt two forms, and are in two colors. Through combination, they can achieve an enriched visual effect in a concise way. The entrance is the top priority in interior design, thus diamond cut shape was mainly applied to the large commercial area to make the entrance a connection between the interior and exterior. Ceiling mixes folded surfaces and oblique lines to present a diamond-like shape and form virtual-real alternated bumping blocks, which adds a sense of quality to the entrance and promotes the grade of the Plaza.

LANDSCAPE DESIGN
广场景观

景观以"壮锦"的编织花纹为元素，提炼出多样而统一的平面铺装图案，结合整体曲线的韵律，展现安吉绿城绿叶繁花之绽放的生命律动。大商业带状广场由造型花坛与大商业建筑围合而成。雕塑《欢聚》、《独秀》造型生动飘逸，热烈活泼，尽显少数民族的热烈与欢腾，视觉效果简洁、大气，将壮族文化展现得淋漓尽致。

The landscape relies on weaved pattern of "Zhuang Brocade" to make a diversified yet unified pavement pattern. Combined with the rhythm of overall curve, it presents the life rhythm of green Anji's green leaves and blooming flowers. The linear plaza of the large commercial area is encircled by shaped parterres and the large commercial buildings. Elegantly and lively shaped sculptures, "Happy Reunion" and "Single Beauty", fully demonstrated the ardency and joy of the minorities. They also demonstrated the culture of Zhuang Nationality with their concise and magnificent visual impression.

10 南宁安吉万达广场喷泉
11 南宁安吉万达广场绿化
12 南宁安吉万达广场主雕塑
13 南宁安吉万达广场室外步行街
14-15 南宁安吉万达广场室外步行街景观小品

OUTDOOR PEDESTRIAN STREET
室外步行街

在金街的立面材料中，运用石材、红砖、铝板及金属格栅构件的错落搭配，营造出丰富多彩的商业氛围。在局部主题店铺设计中采用南宁当地的文化符号，以突出本土建筑的独特性，把大众生活、购物娱乐和历史文化升华为街巷的氛围并转化为营商机遇，为顾客提供创新及充满个性的购物及休闲体验环境。

The facade materials, stones, red bricks, aluminum plates and metal grilles are well-arranged, they build a commercial atmosphere full of variety. The local cultural symbols of Nanning are applied to the design of some theme stores to highlight the uniqueness of local buildings. Meanwhile, such an application successfully transcend public life, entertainment shopping and historical culture into street atmosphere which then turns into commercial opportunities, providing an innovative and featured shopping and leisure environment to the customers.

11 SICHUAN GUANGYUAN WANDA PLAZA
四川广元万达广场

时间：2015 / 06 / 05	OPENED ON : 5th JUNE, 2015
地点：四川 / 广元	LOCATION : GUANGYUAN, SICHUAN PROVINCE
占地面积：12.1公顷	LAND AREA : 12.1 HECTARES
建筑面积：62.7万平方米	FLOOR AREA : 627,000M²

PROJECT OVERVIEW
广场概述

四川广元万达广场位于广元市万源新区，占地12.1公顷，建筑面积62.7万平方米，是广元市最大的商业地产项目。广场包括一站式购物中心、五星级酒店、室内步行街、室外商业街、SOHO公寓、甲级写字楼和住宅等业态。规模巨大的广元万达广场的建设正好切合城市的发展期望，成为集商务、商业、休闲旅游为一体的城市新中心。

Covering a site area of 12.1 hectares and a gross floor area of 627,000 square meters, Guangyuan Wanda Plaza is located in Wanyuan New District, Guangyuan City. The Plaza is the largest commercial real estate project in the city, including several commercial programs, such as a one-stop shopping center, a five-star hotel, an interior pedestrian street, an outdoor commercial street, SOHO apartments, grade-A office buildings and residence. Construction of such a huge plaza lives up to the urban development expectation of the city, and the Plaza is set to be a new city center that brings commerce, business, and leisure tourism together.

01 四川广元万达广场总平面图
02 四川广元万达广场外立面

02

FACADE DESIGN
广场外装

立面设计构思结合当地人文、历史元素和地貌特征，在商业及SOHO的立面设计语言上以"水"为题，用"波"型曲线的元素贯穿始终。临水商业立面设计了一个金色造型，宛如一颗明珠镶嵌在凤冠之上，成为整个临水商业的视觉焦点。在商业主入口也加入的"金色"造型，突出主入口的同时也与主立面造型相呼应。SOHO的立面处理则更强调"竖向线条"，以横向窗间墙的宽窄变化来塑造立面造型，形成大尺度的流畅曲线，打造标志性视觉效果。

Considering the local cultural and historical elements and geomorphic features, the facade design took the theme of "water". It used "wave-like" curve elements for the commercial and SOHO areas. The facade of waterfront commercial area is in gold color, looks like a pearl on a phoenix crown, making it the visual focus of the entire area. The gold color can also be found at main entrance of the commercial area, which makes the entrance stand out but at the same time echo with the main facade. The SOHO facade emphasized "vertical lines". Through size change of transverse wall between windows, the SOHO facade formed a smooth curve in a bigger scale and became visually iconic.

03 四川广元万达广场外立面
04 四川广元万达广场2号入口
05 四川广元万达广场圆中庭
06 四川广元万达广场商业落位图
07 四川广元万达广场室内步行街入口顶棚

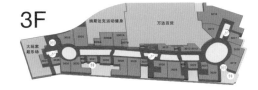

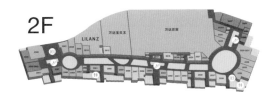

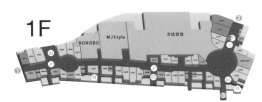

服装　精品　餐饮　体验

INTERIOR DESIGN
广场内装

广场内装以"凤之舞"为设计概念,利用"凤凰"的形体和色彩作为元素,把其婀娜多姿的形态融入设计,构建别具一格、富有文化内涵的购物空间。入口一对凤凰对舞盘旋,形成入口的亮点。门厅在形体延续中求变化,内外呼应,金属材质的应用,增强现代感。圆中庭在大的形体对比基础上进行渗透融合,增加韵律感,有种海纳百川的气质。

Following the idea of "Phoenix Dance", the interior of the Plaza incorporates the graceful figure of phoenix into its design by applying phoenix's shape and color, attempting to render a distinct shopping space full of cultural connotation. The entrance is highlighted by a pair of dancing phoenixes. The lobby seeks variation while keeping the continuation of the form. It looks quite chic through the application of metal materials. The circular atrium tries to blend in while keeping an overall contrast, this adds a sense of rhythm and presents us with an inclusive property.

LANDSCAPE DESIGN
广场景观

广元市种植苍溪雪梨历史悠久，被誉为"中国雪梨之乡"。每年春天，"千树万树梨花开"，美不胜收。广场设计中把白色梨花的静态美发挥到极致，在环境中梨花成为"主角"，形成美好的商业气氛，让人不禁想到梨花香、梨花美的动人景致。

Guangyuan has long planted Changxi Pear and is known as "the hometown of Chinese snow pear". Each spring, the beautiful scene of thousands of pear trees suddenly turn to full blossom can be witnesses. The design maximizes the static beauty of white pear flowers by making them the "main character" in the Plaza, presenting a desirable commercial atmosphere that is closely associated with appealing pear blossom scene and the fragrance.

08 四川广元万达广场主雕塑
09 四川广元万达广场花坛及路灯
10 四川广元万达广场室外步行街
11 四川广元万达广场室外步行街景观小品
12 四川广元万达广场室外步行街

OUTDOOR PEDESTRIAN STREET
室外步行街

室外街的设计结合广元当地悠久的历史文化与优美的基地环境，打造属于广元人民的万达广场，造型交织缠绕、层层交叠，极具特色。通过系列的特色小品，既塑造室外步行街的文化调性、贴近当地文化，营造亲切感和归属感，又为市民喜闻乐见，成为旅游特色，吸引外来人士。

The design of Outdoor Pedestrian Street integrated the long-standing historical culture of Guangyuan and the beautiful site environment, striving to build a Wanda Plaza that is unique to Guanyuan People. By applying a series of special architecture design, the interlaced, overlapped and featured street not only presented a cultural tone, the local culture, the sense of belonging and intimacy, but also pleased the citizens and promoted tourism to attract outsiders.

12 NANTONG GANGZHA WANDA PLAZA
南通港闸万达广场

时间：2015 / 12 / 11
地点：江苏 / 南通
占地面积：12.71 公顷
建筑面积：51.92 万平方米

OPENED ON：11th DECEMBER, 2015
LOCATION：NANTONG, JIANGSU PROVINCE
LAND AREA：12.71 HECTARES
FLOOR AREA：519,200M²

01 南通港闸万达广场鸟瞰图
02 南通港闸万达广场外立面
03 南通港闸万达广场总平面图

PROJECT OVERVIEW
广场概述

广场位于南通市港闸区，占地12.71公顷，总建筑面积51.92万平方米。广场由购物中心、室外商业街、写字楼、公寓及住宅组成。其中购物中心建筑面积17.12万平方米，写字楼建筑面积7.47万平方米，公寓建筑面积6.12万平方米，住宅建筑面积18.39万平方米，其他2.82万平方米。

Nantong Gangzha Wanda Plaza, occupies a site of 12.71 hectares. It is located in Gangzha District of Nantong, Jiangsu Province. It includes a shopping center, an outdoor commercial street, office buildings, apartments and residences. The Plaza has a gross floor area of 519,200 square meters, including 171,200 square meters of shopping center, 74,700 square meters of office building, 61,200 square meters of apartment, 183,900 square meters of residence and 28,200 square meters of other programs.

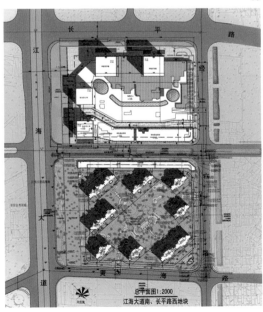

FACADE DESIGN
广场外装

以"扬帆远影"的诗句意向为"起笔",描述万达广场整体规划"以楼为帆"、"以商业为水",犹如巨轮起帆远航的磅礴气势。高层立面以"帆"为基调,彰显挺拔;低层立面以"水中行舟"为源,不失沉稳。"唯见长江天际流"是立面设计的余音,谱写舟与大江融于天际之间、扬起风帆准备远航的华彩乐章。

Starting from Li Bai's verse "lonely sail, distant shadow", the facade design demonstrates a great ship sail-like majestic momentum, through adhering to the overall planning for Wanda Plaza which takes buildings as sail and commerce as water. The facade of higher floors cling to the keynote of "sail" to present towering gesture, and the facade of lower floors is based on the idea of "navigating in water" to ensure a composed gesture. "All I see is the great river flowing into the far horizon" (also Li Bai's verse) is the lingering sound of the facade design, creating a chapter where the boat blend into the skyline, together with the river and sail afar.

05

06

04 南通港闸万达广场外立面
05 南通港闸万达广场椭圆中庭
06 南通港闸万达广场室内步行街连桥
07 南通港闸万达广场商业落位图

07

INTERIOR DESIGN
广场内装

内装设计提取海洋文化元素——海螺、鹦鹉螺、海星的曲线与贝壳的自然纹路和形态，对其解构重组，并结合当地文化设计出具有城市韵味的作品。整体结构以流畅的线条赋予建筑律动的生命气息，并使空间获得扩张感和导向作用。造型的和谐、材料质感的和谐、色调的和谐、风格样式的和谐等，使人们在视觉上、心理上获得宁静、平和的满足感。

The interior design adopts marine culture elements, such as the curves from conch, nautilus and starfish, and the natural texture and shape from shell. The design deconstructed and recombined these elements and added local culture to produce a work with the city charm. The smooth lines of the integral structure endows buildings with energetic life and provide the space with a feel of expansion and guidance. Harmony can be found in modeling, material quality, color and style, which make people visually and psychologically peaceful and satisfied.

08

09

LANDSCAPE DESIGN
广场景观

景观设计提取南通"江海文化"地方风情元素，着力打造一个传承地域特色的商业景观场所。设计围绕"游新濠，赏古景"之主题，将"怡园泊舟，启秀风荷，五亭邀月，仙桥绿堤"等元素融入设计中；同时对建筑整体元素的提炼，将"江海文化"进一步升华，运用于景观铺装、小品及主题雕塑的设计之中，使之与广场建筑产生内在的呼应，"点线面"相互配合，全面映射"江海文化"文脉。

Taking local elements from Nantong's culture of the river and sea, the landscape design strives to build a commercial landscape inheriting regional features. The design centers on the theme of "traveling Hao River, appreciating ancient scenery" and brings elements from the ten sceneries of Hao River into the design. Besides, the architectural elements are extracted to sublimate the culture of the river and sea. They are applied to landscape pavement, landscape pieces and theme sculptures for coherence with buildings. Through the coordination of "points, lines and surfaces", the context of the culture is fully demonstrated.

OUTDOOR PEDESTRIAN STREET
室外步行街

南通室外步行街灵感来源于秦灶老街。追溯到唐朝中叶，南通一带的盐灶以秦灶最为著名，时称通州烧盐第一灶。随着时代的变迁，秦灶老街已经成为南通传统商业文化的一个代表。设计深入挖掘并继承这一鲜明的历史文化特色，把秦灶老街历史元素融入进来，使步行街成为现代而不失传统特色的文化商业街。

The Outdoor Pedestrian Street is inspired by Qin-Stove(today's Qinzao)Ancient Street. Back in the middle of the Tang Dynasty, Qin-Stove was the most famous in Nantong's salt stoves and named as the No.1 salt stove in Tongzhou. As time goes by, Qin-Stove Ancient Street becomes the representation of traditional Nantong commercial culture. Therefore, to dig into and carry on this distinct cultural and historical feature, the design adds the historical element of Qin-Stove Ancient Street, making the Pedestrian Street a cultural and commercial street containing both modern and traditional characteristics.

08 南通港闸万达广场景观
09 南通港闸万达广场主雕塑
10 南通港闸万达广场室外步行街
11-12 南通港闸万达广场室外步行街景观小品
13 南通港闸万达广场室外步行街

13 TAI'AN WANDA PLAZA
泰安万达广场

时间：2015 / 08 / 21　　OPENED ON : 21st AUGUST, 2015
地点：山东 / 泰安　　　LOCATION : TAI'AN, SHANDONG PROVINCE
占地面积：16.21公顷　　LAND AREA : 16.21 HECTARES
建筑面积：65.46万平方米　FLOOR AREA : 654,600M²

PROJECT OVERVIEW
广场概述

泰安万达广场位于泰安市时代发展轴，位于泰山大街以南，灵山大街以北。广场占地16.21公顷，总建筑面积65.46万平方米；其中地上51.46万平方米，地下14.00万平方米，由购物中心、五星级酒店、甲级写字楼、乙级写字楼、SOHO公寓及室外商业街组成。

Located on the Tai'an City development axis, Tai'an Wanda Plaza is to the south of Taishan Street and to the north of Lingshan Street. The Plaza covers a site area of 16.21 hectares and offers a gross floor area of 654,600 square meters, including 514,600 square meters aboveground and 140,000 square meters underground. The overall development consists of shopping center, five-star hotel, grade-A office, grade-B office, SOHO apartment and outdoor commercial street.

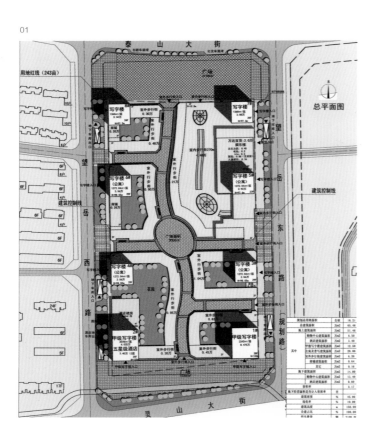
01

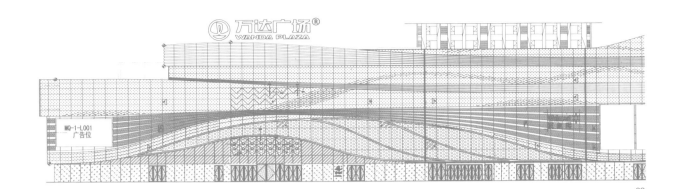

01 泰安万达广场总平面图
02 泰安万达广场立面图
03 泰安万达广场鸟瞰图

FACADE DESIGN
广场外装

建筑特点为高层部分庄严大气并错落有致、低层部分则动感流畅而富于变化。设计从壮观优美的泰山自然风光展开，以"印象泰山"作为设计灵感和契合点——以"塔楼"为山，以"商业"为水。高层立面提取了泰山山脉富有美感的线条作为设计基础元素，大商业部分以水为设计基调，突显了泰山"高山流水"的形象。

The architecture of Wanda Plaza has distinctive character between higher and lower portion of facade. The higher portion is grand and dynamic, and the lower part is flowing and dramatic. Facade design is inspired by spectacular natural scenery of Mount Tai. With the theme of "Impression of Mount Tai", the "tower" mimics the mountain, and the "retail podium" mimics the water. The higher portion facade design takes aesthetic lines of Mount Tai as basic element, while the lower retail podium sets the base as water, together creating an image of "Lofty Mountain and Running Water".

04 泰安万达广场1号入口
05 泰安万达广场外立面
06 泰安万达广场圆中庭
07 泰安万达广场室内步行街入口顶棚
08 泰安万达广场室内步行街电梯
09 泰安万达广场商业落位图

PART **C**   WANDA PLAZAS
万达广场

06

07

08

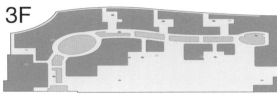

3F

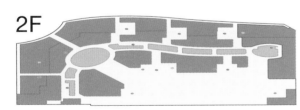

2F

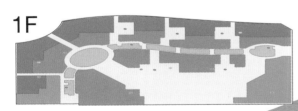

1F

09   1F 时尚名品   2F 潮流体验   3F 休闲餐饮

INTERIOR DESIGN
广场内装

室内步行街着眼于这座城市的地标性景物，以"山水印象"为设计主题，把大商业建筑楼层看作"山"；顶棚提取泰山山脉富有美感的线条作为设计元素，并把该主题运用于地面拼花，将地面看作"水"。椭圆中庭的挑台仿若观光台，成为"赏山观水"的绝佳处所。

Interior Pedestrian Street aims to be the landmark of the city with the theme of "Landscape Impression". The Plaza represents the image of mountain. The aesthetic line of Mountain Tai applies to the ceiling as key design element, which is also applied to ground paving, mimicking the flowing water. The cantilever platform at atrium is the best place for visitors to enjoy the view of mountain and water.

10 泰安万达广场景观
11 泰安万达广场水景
12 泰安万达广场雕塑
13 泰安万达广场室外步行街
14 泰安万达广场室外步行街喷泉

LANDSCAPE DESIGN
广场景观

北广场水景，造型立面略呈"梯形"，与"山"概念寓意一致。水景前高后低，增强形态感，流水朝向大商业，寓意"聚财"。南广场水景，是一个大型镜面水池，池面放置"泰山石海"雕塑。特色花坛组合，造型多为三角和梯形，与地面铺装完美契合。

The water feature at northern plaza adopts the wedge shape from the building facade, with the implication of mountain. The water steps down from front to back to enhance the plaza appearance. Meanwhile, the flowing water towards plaza implies good meaning of "gathering wealth". While the water feature at southern plaza is a large reflection pool, with a sculpture titled "Mount Tai Stones" in the middle. The triangle and wedge shape feature planters are integrated with the ground paving, creating the pleasant landscape environment.

OUTDOOR PEDESTRIAN STREET
室外步行街

金街分为入口段、中央圆形广场段、直段街区和转角节点段等四个分段节点。对不同分段，采用不同的设计元素和手法，为人们带来了不同的建筑体验。设计吸取地域元素，落实"一店一色"的原则，利用多元的设计手法和多样的立面材料，丰富了金街的商业氛围与立面可识别性，打造具有"岱庙"汉代文化特质的休闲类商业街。

The Golden Street consists of four feature blocks, including entrance, central circular plaza, linear block and corner node. Each street block has its own distinctive design element and concept, offering various architectural experience. Integrated with local cultural element, with the design guideline of "One Shop, One Style", the street is designed as a leisure commercial street with the culture character of "Dai Temple". Various facade materials and design technology have been utilised to create a rich commercial atmosphere and enhance the visibility of the architecture.

14 DEZHOU WANDA PLAZA
德州万达广场

时间：2015 / 11 / 06
地点：山东 / 德州
占地面积：16.68公顷
建筑面积：79.38万平方米

OPENED ON: 6th NOVEMBER, 2015
LOCATION: DEZHOU, SHANDONG PROVINCE
LAND AREA: 16.68 HECTARES
FLOOR AREA: 793,800M²

PROJECT OVERVIEW
广场概述

德州万达广场北起新建路，南至新河路，西临城市次主干道湖滨南大道，东至未来的延伸规划路。广场占地16.68公顷，总建筑面积79.38万平方米，包括大型购物中心、室外商业街、写字楼、大型超市和万达影院等，是集购物、餐饮、文化、娱乐等多种功能于一体的大型商圈，也是新的城市中心。

Dezhou Wanda Plaza faces Xinjian Road to the north, Xinhe Road to the south and extended planning road to the east, and borders on Hubin South Road, the urban secondary main road. It has a site area of 16.68 hectares and gross floor area of 793,800 square meters, including shopping center, commercial street, office, large supermarket and Wanda Cinema. The Plaza is set to be a urban complex integrating shopping, catering, culture, entertainment and other functions.

01 德州万达广场总平面图
02 德州万达广场立面图
03 德州万达广场鸟瞰图

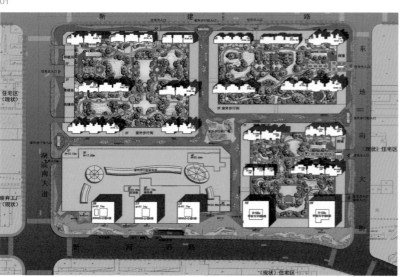

01

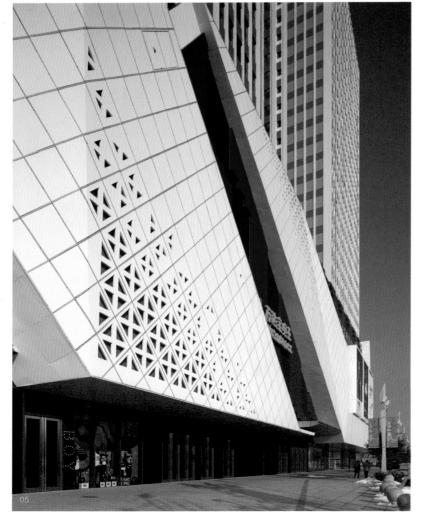

04 德州万达广场外立面
05 德州万达广场外立面特写
06 德州万达广场商业落位图
07 德州万达广场圆中庭

FACADE DESIGN
广场外装

设计理念来源于中国民间艺术中的瑰宝——剪纸艺术的形态特征。建筑立面是仿白纸颜色的铝板，它以带状的形式折叠缠绕，给人带来震撼的视觉。建筑在阳光照射下熠熠生辉，烘托出"巨型剪纸艺术作品"强烈的视觉感受。立面以水平向带状铝板为基本元素，动感而连续，让立面效果既简洁统一又富于变化。入口以白色铝板折叠形成折线状的拱形空间，与剪纸印花玻璃形成强烈对比，成为整个立面里闪耀的"眼"。

The design concept comes from morphological characteristics of the "Paper Cutting", a treasure in Chinese folk art. The paper white color aluminum plate is picked as main facade material. It is folded and twisted in the form of strip, creating dramatic visual impact. The glimmering building under the sunshine is like a piece of "giant paper-cutting works". The basic facade element is horizontal strip aluminum plate, which is dynamic and continuous. The design achieves a simple and unified effect with abundant changes. The folded white aluminum plates frames the main entrance, creating dramatic contrast to paper-cutting paint glass and becoming a shining "eye" of the facade.

INTERIOR DESIGN
广场内装

3F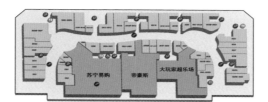

2F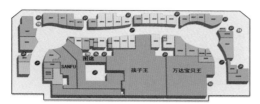

1F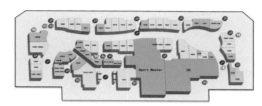

服装 精品 餐饮 体验

内装以"游龙戏水"为设计理念,采用"一黑一白"两种材料,将传说中"龙形龙影"转化为室内设计的语言,追求龙的隐喻效果;同时,造型与人们的流线相呼应,赋予现代商业空间新的生命力。室内步行街以"山东气质"出发,用明确直接的线条和形体来构建空间,并从传统北方剪纸艺术中寻找灵感,界面的牵拉、分离,均具有鲜明的图案美感。

The interior design of Dezhou Wanda Plaza follows the concept of "dragon playing in water". The black and white materials represent the legendary dragon and its shadow, as a metaphor for the concept of dragon. Meanwhile, the well considered pedestrian circulation enhances the quality of retail space. Inspired by the "Essence of Shandong", the design language of straight lines shapes applies to the interior pedestrian street. The idea of traditional northern paper-cutting art has been interpreted to the interior space as feature pattern.

LANDSCAPE DESIGN
广场景观

景观以"水德之兴,舟楫之利"为设计主题,展现德州传统运河文化。体验式景观塑造场景感,打造商业广场的休憩空间;运用当地的人文特色,打造公共艺术空间;结合德州特有的植物,打造具有德州地域特色的植物群落。主入口雕塑《翔》,寓意高飞远翔之意。

The landscape design is themed of "the rise of the virtue of water, the benefit of water transport", which represents the local canal culture in Dezhou. The landscape creates the sense of place, provides pleasant recreational retail space. At the public open space, art pieces have been introduced to showcase the local culture. Local planting species have been selected to form the sense of regional characteristics. The sculpture of "Flying" stands at the main entrance as the symbol of flying high and far.

OUTDOOR PEDESTRIAN STREET
室外步行街

设计挖掘传统艺术，融合当地的文化特色，运用现代建筑设计手法，采用丝网印刷玻璃、塑木、真石漆等材料，创造各具特色又风格统一、传统与时代结合的标准段。各标准段错落有致地组合在一起，形成空间丰富、人文气息浓烈的商业步行街。《欢乐的鱼群》情景雕塑以折纸的形态，表达文化繁荣、经济昌盛的意向。

Typical street block has been designed with the integration of traditional art, local culture and modern design language. Screen printing glass, plastic wood, stone-like paint and other materials have been applied to the street facade to create a unified yet distinctive, traditional yet stylish typical street block. These typical blocks are well arranged to create a retail pedestrian street with various open spaces and cultural characteristics. The sculpture of "Joyful Fish" adopts folding paper form representing the desire for cultural prosperity and economic boom.

08 德州万达广场绿化
09 德州万达广场雕塑
10 德州万达广场座椅
11 德州万达广场室外步行街入口
12 德州万达广场室外步行街
13 德州万达广场夜景

NIGHTSCAPE DESIGN
广场夜景

夜景照明准确把握建筑创意，并与地方文化特色相结合，强化了建筑及地方主题特色；以大商业外立面为照明载体，结合灯光展示当地文化，营造戏剧化的照明效果，打造提升德州城市夜景景观品质的照明盛宴。

The nightscape design aims to enhance the architectural design and local culture. The mall facade is set as the main backdrop, integrated with lighting to present the local culture. A dramatic lighting effect has been created as a lighting feast that enhances the nightscape quality of Dezhou.

15 DONGYING WANDA PLAZA
东营万达广场

时间：2015 / 08 / 15
地点：山东 / 东营
占地面积：17.5 公顷
建筑面积：80.6 万平方米

OPENED ON: 15th AUGUST, 2015
LOCATION: DONGYING, SHANDONG PROVINCE
LAND AREA: 17.5 HECTARES
FLOOR AREA: 806,000M²

PART **C** | WANDA PLAZAS 万达广场 | 167

01 东营万达广场鸟瞰图
02 东营万达广场总平面图

PROJECT OVERVIEW
广场概述

东营万达广场位于山东省东营市东营区，占地17.5公顷，总建筑面积80.6万平方米。广场由购物中心、室外商业街、甲级写字楼、五星级酒店、SOHO公寓及住宅组成。其中购物中心建筑面积15.51万平方米，甲级写字楼建筑面积4.2万平方米，五星级酒店建筑面积3.78万平方米。

Dongying Wanda Plaza is located at Dongying District, Shandong Province. It has a site area of 17.5 hectares and gross floor area of 806,000 square meters. The Plaza consists of shopping center, outdoor commercial street, grade-A office, five-star hotel, SOHO apartment and residential. Among the overall gross floor area, there are 155,100 square meters for shopping center, 42,000 square meters for grade-A office and 37,800 square meters for the five-star hotel.

03

03 东营万达广场入口
04 东营万达广场外立面
05 东营万达广场室内步行街电梯
06 东营万达广场商业落位图
07 东营万达广场入口顶棚

04

FACADE DESIGN
广场外装

大商业建筑将东营市市花"红柳花"进行了抽象化、像素化处理，运用双层釉点玻璃和多孔铝合金面板巧妙地叠加，勾勒出"红柳花"的装饰图案。大商业立面入口区域的流线型设计突出了建筑商业区入口的大方气派，曲线的走势舒缓流畅，在出入口部位构成了显著形象。

The building form of Wanda Plaza is the abstraction of the "Tamarix Flower" (the city flower of Dongying). The facade is applied with double glazed glasses and perforated aluminum alloy plates, with the feature pattern of the flower. The streamline design of plaza entrance area highlights the grandness of the commercial appearance, and offers a distinctive sense of arrival.

INTERIOR DESIGN
广场内装

内装概念是"黄龙入海"。设计对造型加以提炼和简化,以银色金属漆的流动线条为主体,以白色石膏板为衬托,形成上卷、升腾的浪花形态。连桥形体穿插交错,形成跃动不息的节奏。侧裙的元素延续至扶梯,增强水平和纵向贯通性。

To present the idea of "A Yellow Dragon Diving into the Sea", the interior design introduced a series of silver metallic painted lines with the white plasterboard background, creating whirling and rising waves. The interlocking of footbridge forms a dynamic rhythm. Side skirts elements extend towards to escalators, which enhances connectivity in both horizontal and vertical direction.

4F

3F

2F

1F

服装　精品　餐饮　体验

08 东营万达广场雕塑
09 东营万达广场绿化
10 东营万达广场地铺
11 东营万达广场室外步行街
12 东营万达广场室外步行街入口

LANDSCAPE DESIGN
广场景观

景观以"水之魂"与"花之意"为设计主题。"黄河入海"是东营特有的自然景观,黄河入海口的壮丽与长河落日的静美均为难觅的景致。运用抽象化的铺装表现"夕阳西下、长河落日"的美景,带来空间的流动性。主雕塑《黄河入海》也再现了这一大自然的奇观。大商业广场地面衍生建筑外立面装饰肌理,进行细节设计,达成导引人流的效果。

Landscape design takes the theme of "Soul of Water" and "Implication of Flower". Dongying is characterized by the natural scenery of "The Yellow River flowing into sea". The magnificence of yellow river and the tranquil beauty of sunset by the river have been interpreted to the landscape design. The abstract paving pattern shows the beauty of "Sun Setting in the West, Sunset Meeting the River". The sculpture of "The Yellow River Flowing into Sea" represents the marvelous natural wonder. The paving pattern of plaza is responding to the texture of the building facade to direct visitors moving towards the mall.

OUTDOOR PEDESTRIAN STREET
室外步行街

金街将多种风格的店面集中于一体，展现文化的多元性，让不同建筑风格在这里交相辉映，相互碰撞又不失协调统一。让多元文化引领潮流，吸引外来人士到此旅游，给人以深刻印象，实现了"一店一色，层次丰富，传统与现代的结合"的设计初衷。

Showing the cultural diversity, multiple shop front typologies are mixed in the golden street. The different styles are integrated in a harmonious way to attract tourists and local visitors. Thus, the original design intention of "one shop one style, rich gradation, combination of tradition and modern" has been fully implemented.

16 HUANGSHI WANDA PLAZA
黄石万达广场

时间：2015 / 07 / 03
地点：湖北 / 黄石
占地面积：17.14公顷
建筑面积：89.9万平方米

OPENED ON : 3rd JULY, 2015
LOCATION : HUANGSHI, HUBEI PROVINCE
LAND AREA : 17.14 HECTARES
FLOOR AREA : 899,000M²

PROJECT OVERVIEW
广场概述

黄石万达广场位于黄石市黄石港区，占地17.14公顷，总建筑面积约89.9万平方米。广场拥有14.66万平方米大型购物中心、3.25万平方米五星级酒店、7.99万平方米商业街、3.19万平方米国际公寓及8.6万平方米高级写字楼；商业配套体量大、档次高、功能齐全，是黄石高品质城市商业中心。

Located in Huangshigang District, Huangshi City, Huangshi Wanda Plaza covers a site area of 17.14 hectares offers a gross floor area of 899,000 square meters. The development consists of 146,600 square meters shopping center, 32,500 square meters five-star hotel, 79,900 square meters retail pedestrian street, 31,900 square meters luxury apartment and 86,000 square meters high-end office. With such comprehensive mixed-uses, the Plaza serves as high-end urban commercial center of Huangshi.

01 黄石万达广场总平面图
02 黄石万达广场鸟瞰图

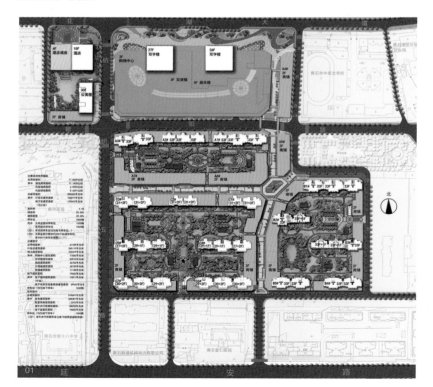

FACADE DESIGN
广场外装

设计取意"湍流不息的长江之水",借用现代主义建筑手法加以表现。大商业立面起伏错落、蜿蜒伸展,黄色波浪墙面与浅灰色墙面相互交织,如江水起伏激荡。穿孔铝板上的孔洞和黄铜色铝板上的交叉网格意为波光。建筑在光影效果的配合下,其灵动性和趣味性一览无遗,体现了商业的活跃氛围。

Facade design takes a modern interpretation to express the idea of "Flowing Yangtze River". The undulating and winding facade of the Plaza represents the image of surging river, with the yellow wavy wall interweaves with light grey wall. Holes on perforated aluminum plates and staggered mesh on brass aluminum plates represents glistening light of waves. Such light & shade effect brings flexibility and enjoyment to activate the building and create a vibrant retail atmosphere.

03

03-04 黄石万达广场外立面
05 黄石万达广场圆中庭
06 黄石万达广场鸟瞰图商业落位图
07 黄石万达广场室内步行街连桥

04

INTERIOR DESIGN
广场内装

室内色彩选择以米色淡雅色系为主色调，整个室内造型和色彩完整、统一、大气，曲线型的动感形态也给人"水城"的联想。在肌理等细节上的处理上也使空间更有层次，表达出黄石的深厚文化底蕴且给人以丰富的想象空间。室内步行街公共空间以凸显当地文化特色为主旨，以贯穿空间的"水元素"表现水的恬静与激情。

Light beige color has been picked to set up the tune of the interior space, bringing the feel of harmony, integration and grandness. Dynamic curve language can be associated with the image of "Water City". The design of texture details helps to build layered space, showcasing the profound local culture. The open space in the interior pedestrian street is designed with water elements, celebrating the city culture.

3F

2F

1F

服装　精品　餐饮　体验

LANDSCAPE DESIGN
广场景观

大商业景观呼应了建筑立面，同时融入了黄石的矿冶文化、自然风貌和青铜文化等元素，营造了雅致而底蕴深厚的景观环境。大商业带状广场是由帆船造型的花坛与大商业建筑围合而成。雕塑以"星河"为表现主题，象征黄石矿冶文化和当地的青山秀水，寓意天地之精华汇聚于此。

The landscape design is responding to the facade design, also adding its own elements such as mining culture, natural landscape and bronze culture of Huangshi, to create an elegant landscape environment with profound culture essence. The linear plaza of Huangshi Wanda Plaza is enclosed by the boat-shaped planter and commercial buildings. The sculpture of "Galaxy" is the symbol of the City's cultural heritage and beautiful nature, a metaphor for gathering essence of nature and heaven in the plaza.

08 黄石万达广场主雕塑
09-10 黄石万达广场景观小品
11-12 黄石万达广场室外步行街

OUTDOOR PEDESTRIAN STREET
室外步行街

金街沿用传统的街道尺度，营造怡人的休闲氛围，体现城市的景观与建筑的风土人情。规划合理的人流动线，提升商业价值。借鉴地标建筑的立意，将不同时代与风格的建筑元素融为一体，发挥黄石的城市魅力，成为当地市民的休闲娱乐与外地游客旅游观光的上选。

The traditional street scale has been applied in the golden street to create a pleasant retail atmosphere, as the showcase of local culture and urban lifestyle. The pedestrian circulation route has been well planned to enhance the retail value. Various architectural elements of different ages and styles have been integrated as a whole to create a distinctive landmark. The Plaza becomes the icon of Huangshi City and a popular destination for both tourists and local residents.

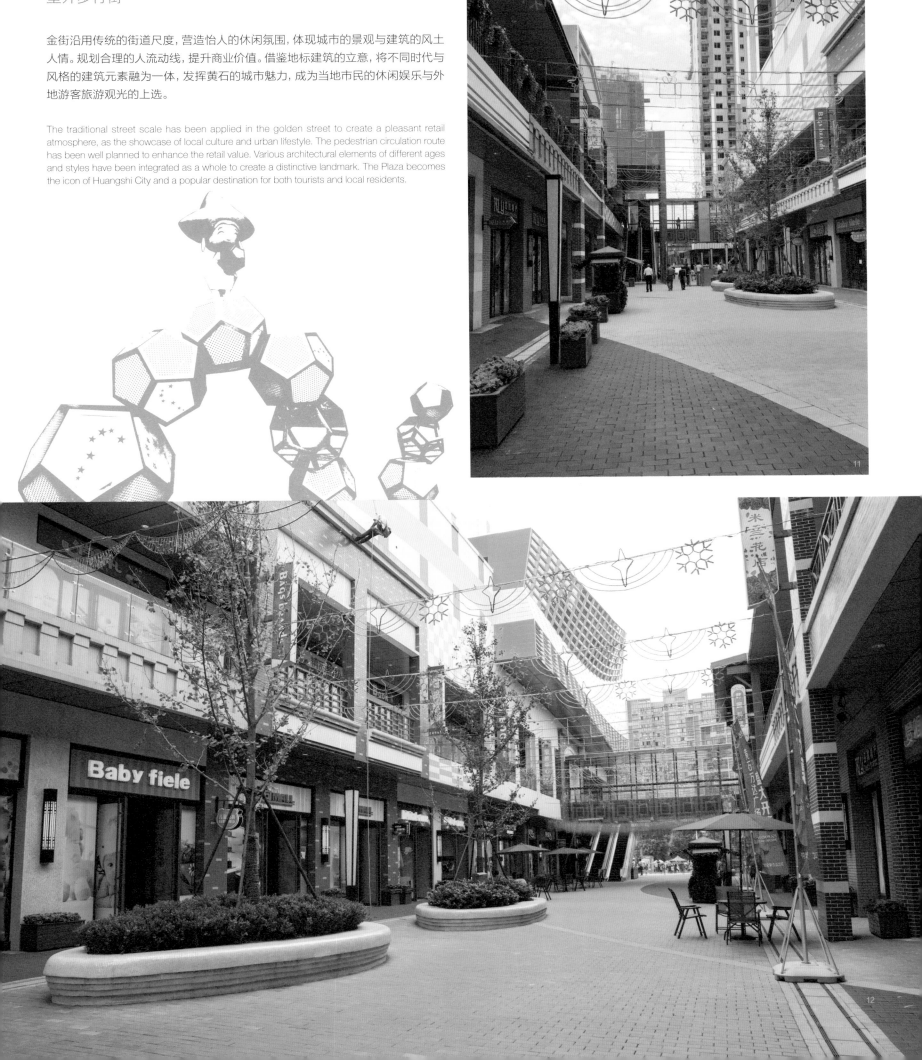

17 ZHEJIANG JIAXING WANDA PLAZA
浙江嘉兴万达广场

时间：2015 / 06 / 12	**OPENED ON**：12th JUNE, 2015
地点：浙江 / 嘉兴	**LOCATION**：JIAXING, ZHEJIANG PROVINCE
占地面积：8 公顷	**LAND AREA**：8.0 HECTARES
建筑面积：31.7 万平方米	**FLOOR AREA**：317,000M²

PROJECT OVERVIEW
广场概述

浙江嘉兴万达广场项目位于浙江省嘉兴市南湖区，占地8.0公顷，总建筑面积31.7万平方米。其中地上部分23.5万平方米，地下部分8.2万平方米，包含购物中心、独立公寓、室外商业街、住宅及底商等业态。

Jiaxing Wanda Plaza is located at Nanhu District, Jiaxing City of Zhejiang Province. It covers a site area of 8.0 hectares and with gross floor area of 317,000 square meters, including 235,000 square meters aboveground and 82,000 square meters underground. The Plaza consists of shopping center, apartment, outdoor pedestrian street stores, and residential with ground floor retail.

01

01 浙江嘉兴万达广场总平面图
02 浙江嘉兴万达广场鸟瞰图

FACADE DESIGN
广场外装

嘉兴素有"丝绸之府"的美誉。设计以嘉兴闻名的丝绸为切入点,以"绸带"作为设计概念。大商业立面采用飘逸的"绸带"形态契合了细长的平面格局,也带来了形态的丰富变化。两条自然卷曲的"绸带"相互叠合与交织,柔美中蕴涵张力,富有视觉冲击力。建筑的天际轮廓也随着卷曲造型而自然起伏,给人印象深刻。

Jiaxing City is well known as the "Hometown of Silk". Thus, based on the city's history, the design introduces the idea of "Silk Ribbon". The graceful ribbon shaped facade of Wanda Mall matches with the linear floor plan layout, at the same time enriches the building form. The two naturally interweaving ribbons bring strong visual impact, showing the tension within peace. The building gives a striking impression with the dynamic skyline spontaneously following with the "Ribbon".

03 浙江嘉兴万达广场外立面
04 浙江嘉兴万达广场入口
05 浙江嘉兴万达广场商业落位图
06 浙江嘉兴万达广场圆中庭
07 浙江嘉兴万达广场室内步行街

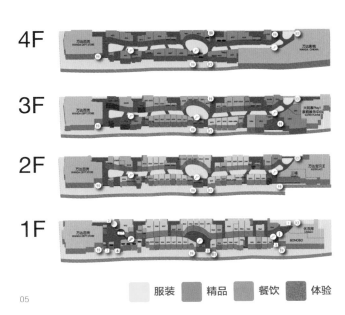

4F
3F
2F
1F

05

服装　精品　餐饮　体验

06

INTERIOR DESIGN
广场内装

内装设计以"典藏历史,蕴含文化,尊重自然"为理念。设计母题为《水漾》,表达"人在水上行,水在城中漾"的意境,体现"古镇——曲水悠扬,宁静生活"、"城市——车水游龙,无限发展"的愿景。设计采用石鱼、玉石器等"马家浜文化"元素,从历史人文中寻找灵感,对商业空间进行文化底蕴丰厚的现代风格构建。

Taken the idea of "collection of history, connotation of culture, and respect for nature", the interior design is themed of "Water Flowing". It expresses the artistic concept of "people walks on water, water meanders in city", delivers the vision of the ancient water town, peaceful life, developing towards a dazzling city image. The feature elements such as stone fish and jade articles abstracted from local Majiabang Culture have been interpreted in the interior space. The design has the inspiration from local history and culture, reconstructing the internal commercial space in modern style.

07

LANDSCAPE DESIGN
广场景观

景观设计以嘉兴特色的"丝绸"为原点，以卷曲"丝带"形成有层次的造型作为景观元素，使得景观卷曲的"实"与建筑物立面玻璃的"虚"相呼应，虚实相间、刚柔并济。材质和细部处理均源于对"丝绸"工艺做法的推敲，具有明显的地域性特征。

Starting from the "Silk" idea, the design language of layered ribbon has been applied in the landscape design. The "solid" landscape has a dramatic contrast to the "void" glass facade, addressing the alternation of virtual and real, and combination of softness and hardness. The materials and finishes of facade design are based on deliberation on silk technique and have distinct regional characteristics.

08 浙江嘉兴万达广场雕塑
09 浙江嘉兴万达广场景观小品
10 浙江嘉兴万达广场景观
11-13 浙江嘉兴万达广场室外步行街

OUTDOOR PEDESTRIAN STREET
室外步行街

将嘉兴当地休闲文化与现代多元文化巧妙融入商业步行街，打造出沿河"酒吧街"及"月老街"两条主题鲜明的地方特色街区。建筑外立面采用"一店一色"的设计原则，结合河流设计景观休息平台、室外休憩空间等，体现步移景异、人在画中游的特点。

The local culture has been skillfully integrated into the outdoor pedestrian street. There are two feature blocks within the street - Bar Street and "Yuelao" (Match-maker) Street. The facade design follows the principle of "One Shop, One Style". Landscape platform and outdoor recreational space are designed along the river, which create various characters of the intimate space.

18 SUZHOU WUZHONG WANDA PLAZA
苏州吴中万达广场

时间：2015 / 12 / 11　　**OPENED ON**：11th DECEMBER, 2015
地点：江苏 / 苏州　　　**LOCATION**：SUZHOU, JIANGSU PROVINCE
占地面积：9.36 公顷　　**LAND AREA**：9.36 HECTARES
建筑面积：49.35 万平方米　**FLOOR AREA**：493,500M²

PART C WANDA PLAZAS
万达广场
185

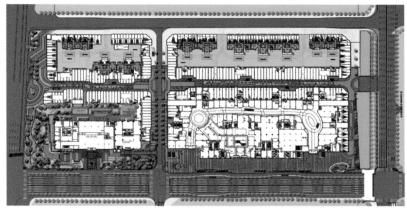

02

PROJECT OVERVIEW
广场概述

苏州吴中万达广场坐落苏州市吴中区石湖西路北侧，旺吴路南侧，友新路东侧，栈廊路西侧。基地位于吴中区的核心区域，地理位置优越，交通便捷，占地9.36公顷，总建筑面积49.35万平方米。广场由购物中心、室外商业街、甲级写字楼、五星级酒店组成。其中购物中心建筑面积16.63万平方米，甲级写字楼建筑面积3.97万平方米，五星级酒店建筑面积3.53万平方米。

Suzhou Wuzhong Wanda Plaza is located at Wuzhong District, Suzhou City, and to the north of Shihu West Road, to the south of Wangwu Road, to the east of Youxin Road and to the west of Zhanlang Road. Being in the core area of Wuzhong District, the site enjoys the superior geographic location and convenient transportation. With site area of 9.36 hectares and gross floor area of 493,500 square meters, the Plaza includes shopping center, outdoor commercial street, grade-A office and five-star hotel. In terms of the mix of gross floor area, it consists of 166,300 square meters commercial area, 39,700 square meters grade-A office and 35,300 square meters five-star hotel.

01 苏州吴中万达广场鸟瞰图
02 苏州吴中万达广场总平面图

03

03 苏州吴中万达广场入口
04 苏州吴中万达广场外立面
05 苏州吴中万达广场商业落位图
06 苏州吴中万达广场室内步行街

04

FACADE DESIGN
广场外装

广场毗邻太湖的石湖景区，以"石湖清波"为概念，层叠、翻卷的铝板营造出水面荡漾的视觉感受，入口处融入拱门、花窗等"苏式"符号塑造出大气优雅又不失细节的总体形象。大商业立面用现代的设计手法，采用弧形线条和垂直线优雅相接，简约、时尚、轻盈，使人轻松愉悦。入口雨篷外挑，恰如空中划过的飘带，吸引人流进入商场，成为强烈的视觉中心。

The Plaza is adjacent to Shihu Scenic District. The concept of "Shihu Wave" has been interpreted in the facade design, using stacked and curled aluminum plates to mimic water ripples. The traditional Suzhou style architectural symbols such as archway and lattice window have been applied at the main entrance, creating the grand sense of arrival. The design language of plaza facade is simple and stylish, with curve and vertical lines interweaving, creating relax and pleasure atmosphere. The overhanging awning at the plaza entrance is like a ribbon flying across the sky. It becomes a strong focal point, attracting the crowds of visitors flowing into the store, and becomes a strong visual center.

INTERIOR DESIGN
广场内装

广场内装采用运用"新中式"的表现手法，内部的色彩遵循"本土建筑"的"灰和白"为主调的构成，穿插了"光的调色"。在空间结构上也提炼了中国传统建筑"檐、斗栱、山墙、窗格"等语言，并运用到室内的"墙、顶、地"当中去，从而构成了一个"顶为天"、"墙为屋"、"地为水"的"室内苏州园林景象"。

The Neo-Chinese style has been implemented to the plaza interior space. Wanda Plaza follows the grey and white color tune taken from local Suzhou architecture, enriched with the color of light. In terms of spatial structure, elements of traditional Chinese architecture-eave, bucket arch, gable and pane are applied to wall, roof and floor. Accordingly, an indoor Suzhou garden has been created, with "roof as the sky", "wall as the building" and "floor as the water".

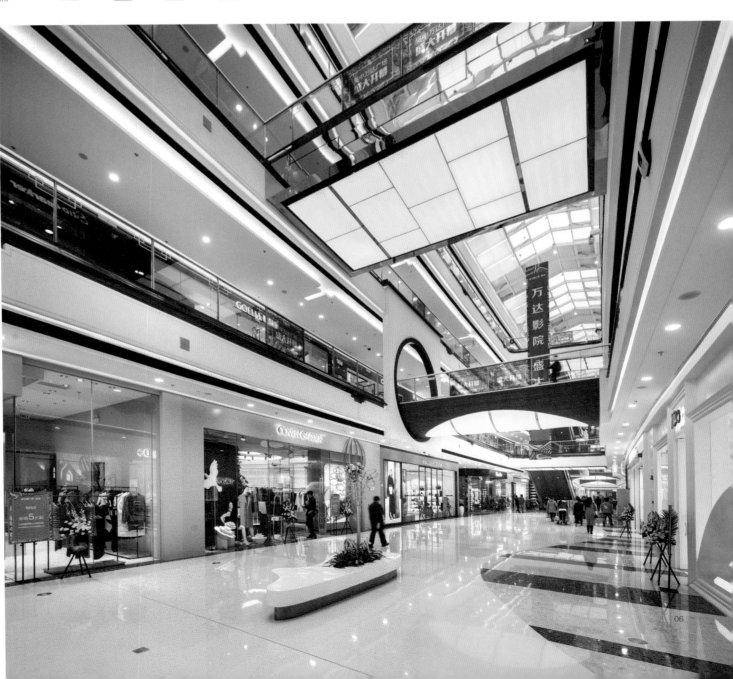

LANDSCAPE DESIGN
广场景观

景观以"吴中水韵"为设计主题，运用动感的曲线表达苏州的河网水系，用具有特定形状的绿化象征水中的太湖石，再搭配具有苏州吴中本土文化特征的景观小品，形成一个具有"吴中水韵"的意境商业景观。以"水"为设计元素，也寓意了财运亨通，流转通达。

To depict the theme of "Wuzhong Water", dynamic curves are introduced in the landscape design to mimic the river network in Suzhou. Sculptural greenscape is designed to symbolize Taihu Lake stone in the water, together with other landscape features, creating an artistic retail environment themed with "Wuzhong Water". Moreover, the "water" element implies good luck in wealth and everything.

07 苏州吴中万达广场景观
08 苏州吴中万达广场主雕塑
09-11 苏州吴中万达广场室外步行街

OUTDOOR PEDESTRIAN STREET
室外步行街

步行街利用现代手法和材料,结合苏州传统建筑特色,以"单跨"或"双跨"为一个单元,创造"一店一色"的独特商业街立面。多种单元之间既独立又有联系,组合方式灵活多样,通过屋顶的高低起伏变化和平面的凹凸进退,达到丰富的空间效果。金街入口门头部分重点处理,引用苏州风雨桥造型,结合二层连廊形成具有地方特色的入口空间,形象简洁大气。

To build a unique commercial streetscape with the character of "One Shop, One Style", the outdoor pedestrian street has been applied with modern techniques and materials. The typology of "single span" or "double span" of traditional Suzhou retail building has been introduced in the street retail. Through the undulating roof and uneven plane, these independent yet interconnected building units can be flexibly combined to achieve rich spatial effect. At the gateway of golden street, the traditional Suzhou wind & rain bridge has been introduced to address the main entrance, also connecting the upper level retail space.

19 FUYANG YINGZHOU WANDA PLAZA
阜阳颍州万达广场

时间：2015 / 08 / 29　　**OPENED ON**: 29th AUGUST, 2015
地点：安徽 / 阜阳　　　**LOCATION**: FUYANG, ANHUI PROVINCE
占地面积：16.13 公顷　　**LAND AREA**: 16.13 HECTARES
建筑面积：69.29 万平方米　**FLOOR AREA**: 692,900M²

PART **C** | WANDA PLAZAS
万达广场 | 191

PROJECT OVERVIEW
广场概述

广场位于阜阳市颍州区，位于中清路以东、河荫路以南、颍州南路以西，淮河路以北，占地16.13公顷，总建筑面积69.29万平方米。项目由购物中心、五星级酒店、室外商业街、公寓及住宅等业态组成。其中购物中心建筑面积约14.8万平方米，五星级酒店3.38万平方米，室外商业街7.5万平方米，公寓9.45万平方米，住宅47.56万平方米。

Fuyang Yingzhou Wanda Plaza is located at Yingzhou District, Fuyang City, and to the east of Zhongqing Road, to the south of Heyin Road, to the west of Yingzhou South Road and to the north of Huaihe Road. With a site area of 16.13 hectares. Among the gross floor area of 692,900 square meters, it consists of 148,000 square meters shopping center, 33,800 square meters five-star hotel, 75,000 square meters outdoor commercial street, 94,500 square meters apartment and 475,600 square meters residence.

01 阜阳颍州万达广场鸟瞰图
02 阜阳颍州万达广场总平面图

FACADE DESIGN
广场外装

建筑为现代主义风格，立面以银灰色为主色调，以横向起伏的线条为主要构图；辅以韵律节奏的收口，衔接起伏变化的印花玻璃，形成间隔变化的构图比例和虚实对比。立面亮点为当地剪纸纹理的印花玻璃，在日光及灯光照明下呈现完全不同的效果，但都紧扣"三清贯颍"的主题，将阜阳斑斓的文化传承下去。

To achieve the modern architectural style, the facade of Fuyang Wanda Plaza appears in interesting proportion with interval change and void-solid contrast, with a basic silver grey color tune, highlighted with dominating horizontal undulating lines, accompanied by rhythmical turns and fluctuant patterned glasses. The paper-cutting patterned glass highlights the building facade, which changes drastically under the sun or lighting. Yet all these effects address the theme of "Three Rivers Running through Yingzhou" and carry on the splendid culture of Fuyang City.

3F

2F

1F

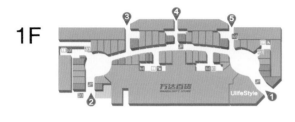

05

03-04 阜阳颍州万达广场外面
05 阜阳颍州万达广场商业落位图
06 阜阳颍州万达广场室内步行街
07 阜阳颍州万达广场椭圆中庭

INTERIOR DESIGN
广场内装

阜阳古时被称为"颍州",如今是安徽省著名的历史文化名城。广场内装秉承了徽派建筑的形式与内涵。商业步行街汲取徽派建筑的特有形式以及当地的人文景观,将提炼的元素运用到重要的人流节点。徽派风格的石雕、木雕、砖雕的运用,结合不同的材质、色彩和质感,从局部到整体都给人协调的美感,带来不一样的视觉享受。

Once called "Yingzhou", Fuyang is now a famous historical and cultural city in Anhui Province. It inherits both form and connotation of "Hui Style" architecture. The retail pedestrian street has the feature elements taken from "Hui Style" architecture and local cultural landscape. The "Hui Style" stone carving, wood carving and brick carving have been applied in interior space with different materials, colors and textures, creating harmonious beauty from details to overall building, and bringing unique visual enjoyment.

LANDSCAPE DESIGN
广场景观

景观以"麦浪"的曲线作为依托,既象征田野里的麦浪,又隐喻颍州西湖的波纹涟漪;同时以麦穗象征丰收的果实,比拟希望的种子茁壮地成长。小品提取阜阳当地的文化,紧扣"春生夏长,秋收冬藏"的主题,主雕塑将蝴蝶与麦穗的造型融合——飞舞的蝴蝶是春天的使者,含羞的麦穗是秋天的果实。

A "Rippling Wheat" curve is introduced in landscape design symbolize ripping wheat in the field and as a metaphor for the ripples in West Lake of Yingzhou. In Chinese culture, wheat is a symbol of harvest and hope. Therefore, the landscape design concept based on the local culture achieves theme of "coming to life in Spring, swelling in Summer, fruiting in Autumn and conserving in Winter". The form of shy wheat and dancing butterfly has been integrated in the plaza sculpture, representing the herald of Spring and harvest in Autumn.

08 阜阳颖州万达广场景观绿化
09 阜阳颖州万达广场导示牌
10 阜阳颖州万达广场室外步行街
11-12 阜阳颖州万达广场室外步行街景观小品

OUTDOOR PEDESTRIAN STREET
室外步行街

将阜阳当地传统文化与徽派文化巧妙融入商业步行街，缔造出"文商并重，古今融合"的气息，打造一条徽派风情与当地文化历史相结合的特色街。通过一系列的特色小品，塑造室外步行街贴近当地文化、具有亲切感和归属感的调性，又为民众和游客增添喜闻乐见的话题。

The local tradition and Anhui-style culture are well integrated into the outdoor pedestrian street. A feature Anhui-style Street has been built to address the atmosphere of "the equal of culture and business, the merging of the past and the present". A series of landscape features have been placed in the street, creating the intimacy and sense of belonging. The outdoor pedestrian street not only presents local culture, but also becomes a hot topic by the public and tourists.

20 NEIJIANG WANDA PLAZA
内江万达广场

时间：2015 / 06 / 26
地点：四川 / 内江
占地面积：15.4公顷
建筑面积：75.2万平方米

OPENED ON: 26th JUNE, 2015
LOCATION: NEIJIANG, SICHUAN PROVINCE
LAND AREA: 15.4 HECTARES
FLOOR AREA: 752,000M²

PROJECT OVERVIEW
广场概述

广场位于四川省内江市东兴区汉安大道以北，国道321线以南、省道206线以东，占地15.4公顷，总建筑面积75.2万平方米。广场由购物中心、室外商业街、甲级写字楼、五星级酒店、公寓和住宅组成；其中购物中心建筑面积14.05万平方米，甲级写字楼建筑面积5.32万平方米，五星级酒店建筑面积3.26万平方米，室外商业街8.0万平方米。

Neijiang Wanda Plaza is located at Dongxing District, Neijiang City, and to the north of Hanan Road, to the south of China National Highway 321, to the east of Provincial Road 206. The site area is 15.4 hectares and with gross floor area of 752,000 square meters, including 140,500 square meters for Wanda Mall, 53,200 square meters for grade-A office, 32,600 square meters for five-star hotel and 80,000 square meters for outdoor commercial street.

03

01 内江万达广场总平面图
02 内江万达广场鸟瞰图
03 内江万达广场外立面
04 内江万达广场入口

04

FACADE DESIGN
广场外装

大商业外立面结合四川本地特色，以"石"、"船"、"山"为元素，将万达气场与地方文化有机融合；细节设计丰富多变，注重"船体"塑造，蕴含节奏感与序列感。立面材料为透明玻璃、彩釉玻璃和铝板，具有简约、时尚与轻盈之感。外挑的铝板造型，形成磅礴的体量与恢宏的气场，成为强烈的视觉中心，是内江区域当之无愧的地标建筑。

The local Sichuan features, such as "Stone, Boat and Mountain" have been integrated to the facade design of Wanda Mall. The boat shape element has been well crafted in facade detail design, showing the sense of rhythm and sequence. Transparent glass, enameled glass and aluminum plate are selected as the facade materials, creating a simply and stylish appearance. Overhung aluminum plate gives a powerful visual impact, which makes the Plaza a well-deserved landmark of Neijiang City.

INTERIOR DESIGN
广场内装

广场内装巧妙地将"川西民居"引入室内,采用含蓄的表达方式表达地方风貌。室内步行街以一条流线统领全局。椭圆厅侧板采用四川特有的屋脊元素,使整体空间环境清晰明快;采用暖色光源衬托较为浓郁的商业氛围。圆厅以丝带及波浪线条作为地面及侧板的装饰元素,结合圆厅较强的空间感,创造高级购物环境的可信赖感。

The concept of "Western Sichuan Folk House" has been implemented in the interior design of Neijiang Wanda Plaza, celebrating the local culture in a subtle way. The interior pedestrian street connects the whole interior space. The Sichuan style pitch roof has been designed on the side plates of elliptical hall, providing a pleasant shopping environment. Warm color lighting sets up a welcoming atmosphere of the Plaza. In the circular hall, a design language of ribbons and wavy lines applies to the ground paving and side plates, which creates the high-end spatial quality.

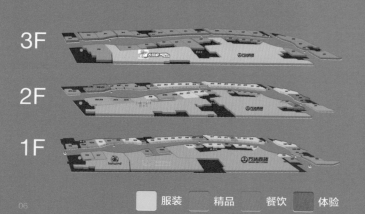

3F / 2F / 1F

服装　精品　餐饮　体验

05 内江万达广场椭圆中庭
06 内江万达广场商业落位图
07-08 内江万达广场室外步行街景观小品
09 内江万达广场室外步行街

OUTDOOR PEDESTRIAN STREET
室外步行街

内江"新四方块步行街"设计注重将内江当地休闲文化与大千山水巧妙融合。步行街既是具有当地历史文化的特色街，又是注重与时俱进的时尚街。内江"新四方块"步行街采用简欧的建筑风格，注重与内江"老四方块街"之间的差异和联系，与其相互辉映，形成双核，与购物中心室内步行街和其他业态互补，相辅相成，成为内江的城市新中心。

The design of Neijiang outdoor pedestrian street achieves the integration of local culture and the essence of Chang Dai-Chien's Painting. It is a place not only full of local history and culture, but also with trendy lifestyle. The "new square pedestrian street" is in simple European style, creating a visual contrast with the original square pedestrian street. The combination of old and new offers double retail cores complementing to the Wanda Plaza, becoming the new city center of Neijiang.

21 QIQIHAR WANDA PLAZA
齐齐哈尔万达广场

时间：2015 / 08 / 01
地点：黑龙江 / 齐齐哈尔
占地面积：12.5 公顷
建筑面积：55.4 万平方米

OPENED ON: 1st AUGUST, 2015
LOCATION: QIQIHAR, HEILONGJIANG PROVINCE
LAND AREA: 12.5 HECTARES
FLOOR AREA: 554,000M²

PROJECT OVERVIEW
广场概述

齐齐哈尔万达广场位于黑龙江省齐齐哈尔市建华区,占地面积12.5公顷,总建筑面积55.4万平方米,场区分为南、北两个地块。南区由购物中心、五星级酒店、SOHO公寓及商铺组成;北区由住宅及配套公建组成。购物中心总建筑面积14.93万平方米,其中地上建筑面积8.58万平方米,地下建筑面积6.35万平方米。

Qiqihar Wanda Plaza is located at Jianhua District, Qiqihar City of Heilongjiang Province, covering a site area of 12.5 hectares and with gross floor area of 554,000 square meters. The Plaza is divided into south and north part, including shopping center, five-star hotel, SOHO apartment and retail. The north part consists of residential and public amenities. The shopping center of the development offers a gross floor area of 149,300 square meters, including 85,800 square meters aboveground and 63,500 square meters underground.

01 齐齐哈尔万达广场鸟瞰图
02 齐齐哈尔万达广场立面图
03 齐齐哈尔万达广场总平面图

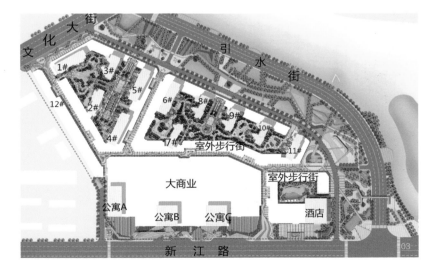

04 齐齐哈尔万达广场外立面特写
05 齐齐哈尔万达广场外立面

FACADE DESIGN
广场外装

齐齐哈尔所处的地理位置给它带来了大自然的独特礼物——冰雪。齐齐哈尔人化严寒为神奇，赋冰雪以生命，用冰雕艺术装扮北国的冬天。"冰凌"成为广场大商业的立意主题——冰凌堆积形成如同河川般的壮美景观。建筑采用极具力量感的折线型体块进行穿插组合，以玻璃作为背景，取得晶莹、纯粹的美学效果。

Due to its location, Qiqihar receives unique gift from nature - Ice and Snow. The severe cold weather with ice and snow has been given magic life by Qiqihar people. Ice sculptures are made to decorate the freezing winter in northern China. "Icicle" is the theme of Qiqihar Wanda Plaza, which represents a grand picture of spectacular glacier. The powerful polygonal building blocks are interlocked together with the glass background, presenting the pure beauty of the architecture.

06

07

08

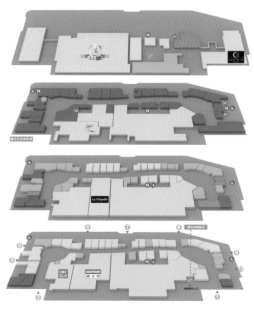

09

 服装　 精品　 餐饮　 体验

INTERIOR DESIGN
广场内装

内装以"仙鹤独立"为设计主题。蔚蓝的天空、广袤的大地,使这片土地的空间感得以无限延展。通过流畅的空间组织、新鲜的建筑材料、温暖而又变化多端的灯光,表现万达广场作为"鹤城"齐齐哈尔首屈一指的建筑形象。设计结合本地居民生活习惯,打造一座艺术与商业完美融合的商场,让其尽情地在这座城市璀璨绽放。

Themed of the "Standing Crane", the blue sky and the vast land within the interior of Qiqihar Wanda Plaza visually extend the space to infinity. The design strategies such as the smooth spatial organization, new technology of building materials, dynamic lighting effects, have been implemented to make Qiqihar Wanda Plaza the landmark of the city. A shopping mall integrated with art, local culture and lifestyle becomes the attractive destination in the city.

06 齐齐哈尔万达广场圆中庭
07 齐齐哈尔万达广场入口顶棚
08 齐齐哈尔万达广场室内步行街连桥
09 齐齐哈尔万达广场商业落位图
10 齐齐哈尔万达广场椭圆中庭

LANDSCAPE DESIGN
广场景观

把建筑外立面元素——浮冰、倒影、绿岛——加以提炼运用到景观设计中,使建筑与景观融合在一起。在水中屹立的"丹顶鹤"造型,象征着万达广场的远大前程。商业街筛选提炼"丹顶鹤"的形象作为设计元素,于是"丹顶鹤"作为整个故事的"主角"以拟人化、动漫化、趣味化的形象出现,打造一条富有人情味的主题步行街。

The facade design elements such as Floating Ice, Reflection and Green Island have been reinterpreted to landscape design, bringing the architecture and landscape as a whole. The image of "Red-crowned Crane" symbolizes brilliant prospects of the Plaza. This "Crane" image becomes the key design element applying to the whole retail street. The character of "Red-crowned Crane" brings fun and characteristics to activate the pedestrian street.

11 齐齐哈尔万达广场雕塑
12 齐齐哈尔万达广场景观
13 齐齐哈尔万达广场室外步行街
14 齐齐哈尔万达广场室外步行街景观小品

OUTDOOR PEDESTRIAN STREET
室外步行街

步行街以"鹤舞"为立意主题，强调齐齐哈尔特有的"鹤文化"。在若干重要视觉节点，比如湖滨广场入口、扶梯外罩及部分门头的外立面，设计采用将红褐色方形铝管变截面或者采用大量圆形小像素灯组合的手法，打造出丹顶鹤在空中飞舞的抽象形体，使整条金街在满足 "一店一色"要求的前提下，既具有强烈的寓意，又增添无限的动感。

Pedestrian Street is theme of "Crane Dance" addressing the "Crane Culture" of Qiqihar. A bespoke lighting feature mimicking a dancing crane has been placed at the key nodes along the street, such as lakeside plaza entrance, escalator cover and certain gateways. Based on the design requirement of "One Shop, One Style", the character of dancing crane brings strong identity to the pedestrian street and animates the streetscape.

22 ANYANG WANDA PLAZA
安阳万达广场

时间：2015 / 07 / 25　**OPENED ON**: 25th JULY, 2015
地点：河南 / 安阳　　**LOCATION**: ANYANG, HENAN PROVINCE
占地面积：11.2公顷　**LAND AREA**: 11.2 HECTARES
建筑面积：64.34万平方米　**FLOOR AREA**: 643,400M²

PROJECT OVERVIEW
广场概述

安阳万达广场位于安阳市文峰区中华路以东、迎春东街以南、永明路以西，占地面积11.2公顷，总建筑面积64.34万平方米，为集大型购物中心、五星级酒店、高档住宅、精装公寓、写字楼和室外商业街等为一体的大型城市综合体，是安阳新的城市中心和经济"助力器"。

Anyang Wanda Plaza is located in Wenfeng District, Anyang City, to the east of Zhonghua Road, to the south of Yingchun East Road and to the west of Yongming Road. With site area of 11.2 hectares and gross floor area of 643,400 square meters, the Plaza is a large-scale urban complex including large shopping center, five-star hotel, high-end residence, luxury apartment, office and outdoor commercial street. It becomes the new city center and economic "accelerator" of Anyang City.

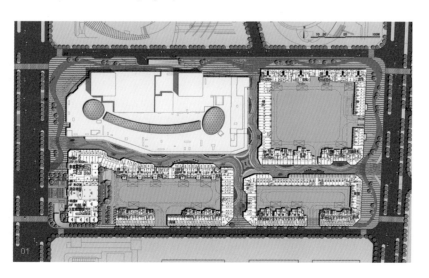

01 安阳万达广场总平面图
02 安阳万达广场鸟瞰图
03 安阳万达广场1号入口

FACADE DESIGN
广场外装

广场外立面造型取安阳的"太极两仪"概念，体块互相咬合，宛然一体，红白双色的波浪墙面使整个商业立面极具流动感、活跃性及现代气息，流畅且富有韵律。圆孔和方空的艺术穿孔板穿插在波纹状的墙面上，增强了整体的动感，活跃了商业氛围；用抽象的波浪形成的巨大拱形彰显商场入口的趣味性、识别性和导入性，以吸引人流的注意。

The facade design concept comes from the "Yin" and "Yang" of Tai Chi. The interlocking blocks forms the complete shape. Red-white corrugated facade elements create the dynamic and contemporary characteristics. The artistic style perforated plates integrate with corrugated walls activating the commercial atmosphere. A giant arch of abstract wave frames the main entrance of Wanda Mall, addressing its identity and visibility.

04

04 安阳万达广场2号入口
05 安阳万达广场外立面
06 安阳万达广场商业落位图
07 安阳万达广场椭圆中庭
08 安阳万达广场室内步行街

05

4F
3F
2F
1F

服装　精品　餐饮　体验

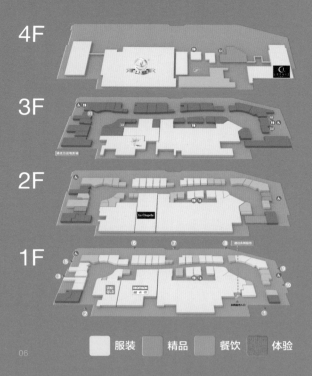

INTERIOR DESIGN
广场内装

广场内装的设计主题是"地域文化的时尚诠释",从型体和材质方面加以"扣题"地表达。形体——提取安阳人文及地域文化的符号语言,营造具有现代感、地域感、特色感、形态的独特商业空间氛围。材质——采用纹理造型GRC与白色乳胶漆及铝板、镜钢,突出地域文化与时尚的融合。

The theme of "Fashion Interpretation of Regional Culture" has been applied into interior design of Anyang Wanda Plaza. The design approach is expressed by shape and materials. In terms of shape, symbolic elements of Anyang local culture has been reinterpreted in the interior space, creating the contemporary retail atmosphere with regional context. Carefully selected materials, such as texture-shaped GRC, white emulsion paint, aluminum plate and mirror steel, have been applied to the plaza interior space, which enhance the integration of local culture and global fashion.

LANDSCAPE DESIGN
广场景观

广场景观的设计主题为"洹河印记"。在形体塑造方面，提取建筑元素，产生河流相互"咬合"的效果，形成曲线现代元素。在色彩塑造方面：以鲜艳的色调与建筑协调一致，使景观与建筑融为一体。

Anyang Wanda Plaza's landscape has been themed "Huan River Impression". The form of landscape is corresponding to the architectural shape. An elegant curve has been introduced to the plaza mimicking the river. In terms of color selection, bright colors are applied, echoing the color tune of buildings and offering a subtle blend of landscape with architecture.

09 安阳万达广场路灯
10 安阳万达广场花坛
11 安阳万达广场景观
12 安阳万达广场室外步行街路灯
13 安阳万达广场室外步行街入口
14 安阳万达广场室外步行街夜景
15 安阳万达广场室外步行街

OUTDOOR PEDESTRIAN STREET
室外步行街

室外步行街的设计主题为"热闹都市",以此表达广场在安阳的商业地位。在形体塑造方面,通过"中西"元素的结合、各种材质的混搭、造型各异的细节及丰富的美陈,营造热闹繁华的都市街道场景,达到设计目的。室外步行街所采用的红色面砖、暖色涂料以及暖色格栅达到色彩和谐统一,也帮助设计目标的实现。

The design of outdoor pedestrian street achieves the theme of "Bustling City", which is showcasing the critical positioning of Anyang Wanda Plaza. The mix of Chinese and Western design elements, variety range of materials, and rich details and colors create a bustling urban street scenery. Meanwhile, the pleasant streetscape has been enhanced by the harmonious combination of red face bricks, warm color coatings and warm color grilles.

23 WEINAN WANDA PLAZA
渭南万达广场

时间：2015 / 06 / 07
地点：陕西 / 渭南
占地面积：6.9公顷
建筑面积：29.0万平方米

OPENED ON: 7th JUNE, 2015
LOCATION: WEINAN, SHAANXI PROVINCE
LAND AREA: 6.9 HECTARES
FLOOR AREA: 290,000M²

PROJECT OVERVIEW
广场概述

渭南万达广场项目坐落于渭南市高新区，位于新区东路以西、东兴街以北、敬贤大街以南，占地6.9公顷，总建筑面积29.0万平方米。广场由购物中心、室外商业街、写字楼、公寓以及五星级酒店等构成。其中购物中心建筑面积7.1万平方米，写字楼建筑面积4.06万平方米，公寓6.0万平方米，室外商业街3.2万平方米。

Weinan Wanda Plaza is located at Weinan Hi-tech Zone, and to the west of Xinqu East Road, to the north of Dongxing Street and to the south of Jingxian Street. The site area is 6.9 hectares, with a gross floor area of 290,000 square meters, including 71,000 square meter shopping center, 40,600 square meters office, 60,000 square meters apartment and 32,000 square meters outdoor commercial street.

01 渭南万达广场总平面图
02 渭南万达广场鸟瞰图

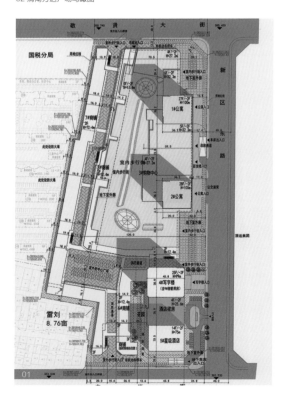

FACADE DESIGN
广场外装

大商业立面以陕西"渭河平原"为"题"、当地风土人情为"材",开始设计之旅。形态上,将渭河地貌之恢宏大气的意境融入空间中,呈现宛如"渭水横贯八百里秦川"的壮丽画面;色调上,以关中平原大地沉稳的色彩为"墨盒",局部穿插民族特色的抽象色体系。建筑外观具有刚性之美,表达关中人民延绵千年的生命活力。

The design concept takes "Weihe Plain" as its "theme" and local culture as its "material". The grandness of Weihe River landform has been interpreted into the facade shape, as presenting a magnificent picture of "Weihe River Spanning Guanzhong Plain". In terms of color selection, the dignified Guanzhong Plain color is served as "ink box", which is interwoven with abstract color series from local culture. The building is powerful in appearance, representing the vitality of Guanzhong people.

INTERIOR DESIGN
广场内装

渭南万达广场的内装设计理念延续了外装的"关中山水"概念，力求避免过度装饰，着重强调内外一体，浑然天成，将水流弧线的元素融会贯穿整个室内步行街。在细节处理方面，注重采用体现当地文脉的元素，表达万达广场"根植关中，服务渭南"的决心。

The interior design inherits the idea of "Guanzhong Landscape". The design approach is to address the integration of interior and exterior, avoiding over design. A simple curve element seamlessly connects the whole interior pedestrian street. Local context elements have been interpreted into many interior details, expressing the Plaza's commission of "Rooted Guanzhong, Serving Weinan".

LANDSCAPE DESIGN
广场景观

建筑的主题是"山茶花"。景观契合建筑的主题,使景观与建筑形成内在的协调。景观采用"生长的叶"的主题,通过立体三维景观的塑造,传递"花与叶"的关系;"生长的叶"也体现在地面铺装、花坛、树池以及景观小品与植物上。通过多层次的立体打造,"花与叶"呈现出全方位的再现。

"Camellia" is the theme of the architecture. Corresponding to this theme, the landscape is well integrated with architecture. The three-dimensional landscape of "Growing Leaf" conveys the relationship of "Flower and Leaf". Meanwhile, the concept applies to every detail such as ground paving, planter, tree pit and other landscape features. The "Flower and Leaf" theme has been fully presented through landscape design in multi levels.

03 渭南万达广场椭圆中庭
04 渭南万达广场室内步行街
05 渭南万达广场绿化
06 渭南万达广场室外步行街

OUTDOOR PEDESTRIAN STREET
室外步行街

将渭南当地的地域文化与新古典主义的现代商业街巧妙结合,打造出一条尺度宜人、氛围融洽的休闲步行街。步行街之中又有一广场置于其中,人们在这里相遇、流连,四周的小店里传来欢声笑语。在广场中央举行各种活动吸引周边的百姓争相涌来,此处俨然成为渭南新的休闲中心。

The Weinan local culture has been blended into the neoclassic style retail street, offering a pleasant human scale pedestrian street. A plaza is located in the middle of the street, for people gathering, hanging around, also as active open space for retail promotion attracting the crowds. The street has thus become the new leisure attraction of Weinan City.

24 YINGKOU WANDA PLAZA
营口万达广场

时间：2015 / 06 / 19
地点：辽宁 / 营口
占地面积：13.06公顷
建筑面积：67.27万平方米

OPENED ON : 19th JUNE, 2015
LOCATION : YINGKOU, LIAONING PROVINCE
LAND AREA : 13.06 HECTARES
FLOOR AREA : 672,700M²

PROJECT OVERVIEW
广场概述

营口万达广场位于营口市体育馆路东侧、市府路西侧、渤海大街南侧、体育场南路北侧，总占地面积13.06公顷，建筑面积67.27万平方米。作为城市综合体，广场包括大型商业中心、城市步行街、写字楼、公寓和住宅等，集购物、餐饮、文化、居住、办公和娱乐等多种功能于一体。其中，购物中心建筑面积17.02万平方米，销售物业建筑面积50.25万平方米。

Yingkou Wanda Plaza is located in Yingkou City, and to the east of Tiyuguan Road, to the west of Shifu Road, to the south of Bohai Street and to the north of Tiyuchang South Road. The site area is 13.06 hectares, with a gross floor area of 672,700 square meters, including 170,200 square meters of shopping center and 502,500 square meters for sale. Being an urban complex, the Plaza comprises of large shopping center, pedestrian street, office, apartment and residential, as a mixed-use urban development.

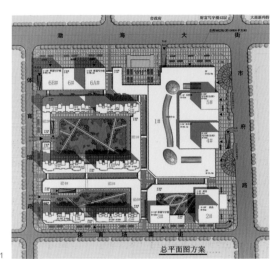

01 总平面图方案

01 营口万达广场总平面图
02 营口万达广场2号入口
03 营口万达广场外立面
04 营口万达广场立面图

FACADE DESIGN
广场外装

设计构思来源于营口这座海边城市的海边岩石形态，通过宛如两块叠落"岩石"的处理来形成立面的风姿。通过"岩石"表面的转折和切削来形成硬朗而有层次的外观，而穿插于岩石体块之间象征水的玻璃体量又柔化了整体效果，使得立面刚柔并济，浑然一体。入口采用岩石般的铝板悬挑，结合竖向的玻璃，使得大商业的入口显得轻盈迷人。

The design inspiration comes from the rocks form along the shoreline of Yingkou. The facade looks like two stacked "rocks", the turning and cutting of which contribute to a layered appearance. Meanwhile, the glass between the two rocks symbolizing water intersperse rocks, adding softness to balance, making facade an integrated whole. The rock-shape overhung aluminum plates frames the main entrance, integrated with vertical glasses, creates a lightweight and enchanting effect.

INTERIOR DESIGN
广场内装

室内步行街以"点、线、面"全方位的整体构思来塑造室内商业空间,采用了"穿插"和"加减"的设计手法使得空间丰富、引人入胜。装饰简洁明快,不同功能的空间虽然大小有别,但总体色调和设计风格把控得统一协调,简洁中透露着大气。

The design of interior pedestrian street brings the idea of holistic "point, line and surface" to create interesting commercial interior space. By inserting and subtracting elements, it enhances richness of the space. The overall interior decoration is in simple and clean style, with various spaces in different size and functions. It achieves the harmony of overall color and design style, showing the charm of grandness in simplicity.

LANDSCAPE DESIGN
广场景观

景观设计以沙滩上的"贝壳"为理念缘起,确定了海滩上"海浪痕迹",海潮退去后留下"海中瑰宝"等主题,形成"海之趣"、"海之汇"和"海之浪漫"的"三大海洋文化景观节点",在营造"宜人尺度"空间的同时,也连贯了步行街内的景观空间。

Inspired by the "Shell" on the beach, the landscape design concept is of "Wave Trace" on the beach and "Sea Treasure" when the tide goes out. It creates three landscape nodes of marine culture, including "Haizhiqu (Interesting Sea)", "Haizhihui (Scenic Sea)" and "Haizhilangman (Romantic Sea)". The landscape space creates a pleasant human scale environment and connects to pedestrian street.

05 营口万达广场圆中庭
06 营口万达广场入口天花
07 营口万达广场景观雕塑
08 营口万达广场室外步行街

OUTDOOR PEDESTRIAN STREET
室外步行街

步行街色调以暖色调为主,局部加以冷色调点缀;同色系暖色调颜色的变化,营造丰富的商业氛围;局部冷色调的运用,活泼了立面形态。"吉祥鱼宝"雕塑设计紧密围绕"海潮汇"这一景观主题,意图鲜明、造型新颖别致。

The color theme of pedestrian street is based on warm color tune and highlighted by cold tune. The variation within warm color tune create colorful commercial atmosphere. The facade is partly highlighted by cold tune color, which activates the whole street. The sculpture "Auspicious Fish" is the most distinctive icon in the street, representing the theme of "Tide Confluence".

25 JIAMUSI WANDA PLAZA
佳木斯万达广场

时间：2015 / 09 / 12
地点：黑龙江 / 佳木斯
占地面积：27.8公顷
建筑面积：105万平方米

OPENED ON : 12th SEPTEMBER, 2015
LOCATION : JIAMUSI, HEILONGJIANG PROVINCE
LAND AREA : 27.8 HECTARES
FLOOR AREA : 1,050,000M²

PROJECT OVERVIEW
广场概述

佳木斯万达广场位于佳木斯郊区光复路以南、万新街以西、八一街以东，占地面积27.8公顷，总建筑面积105万平方米，为集大型购物中心、五星级酒店、写字楼、高档住宅、城市SOHO和室外商业街区为一体的大型城市综合体。

Jiamusi Wanda Plaza is located in the suburb of Jiamusi City. The site is to the south of Guangfu Road, to the west of Wanxin Street and to the east of Bayi Street. With the site area of 27.8 hectares and gross floor area of 1,050,000 square meters, the Plaza is a large-scale urban complex with mixed-use of large shopping center, five-star hotel, office, high-end residential, urban SOHO and outdoor commercial street.

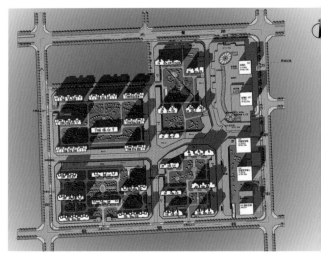

01

01 佳木斯万达广场总平面图
02-03 佳木斯万达广场外立面
04 佳木斯万达广场1号入口

FACADE DESIGN
广场外装

设计遵循整体统一的原则,采用简洁时尚的风格,注重提取地方特色符号加以现代应用——如将江河的形态、赫哲族鱼皮服饰的元素等完美融合到整体和细部的造型中;以不同质感和颜色细微变化的铝合金幕墙模拟鱼皮纹饰,取得随着光影的变化而变幻的效果。抽象鱼皮服饰的色彩与肌理作为外立面的元素,使得建筑具有生命力。

The facade design follows the principal of unity and adopts simple & chic style. It has absorbed local cultural symbols and applied with modern touch. The shape of river and the unique fish skin pattern of Hezhe Nationality dress have been integrated into the overall and detail design. The different textures and subtle changes in color of aluminum alloy curtain wall mimics fish skin dress pattern along with light & shadow. The abstract fish skin dress pattern has been applied on facade, which brings the vitality to the building.

INTERIOR DESIGN
广场内装

内装彩色跳跃式的线框造型点缀在白色的空间之上,活化了商业气氛。顶棚与地面相呼应,用色块拼贴的方式表现空间的灵动性。立面造型和色彩与街区空间和谐相处,线性的设计具有冲击感及延展性。地面的设计手法彰显个性,渐变式的拼贴效果带来别样的视觉感受。

The interior of Jiamusi Wanda Plaza is white space decorated with colorful wireframe pattern, which is activating the commercial atmosphere. The ceiling and floor echo each other, with color patch collage to reinforce the spatial flexibility. The facade design matches with the internal street space. The linear pattern creates strong visual impact. The gradual changing collage pattern of floor design gives unique personality of the interior space.

05 佳木斯万达广场入口内装
06 佳木斯万达广场椭圆中庭
07 佳木斯万达广场景观绿化
08 佳木斯万达广场室外步行街

LANDSCAPE DESIGN
广场景观

景观着重诠释"三江平原"星罗棋布的岛屿与金色河流的富饶美丽；三角形石材运用于大商业主入口处，强化了"东极之光"的璀璨之感，塑造了引人入胜的入口空间。岛状花坛以极具张力的面貌呈现于广场，选用北方特有植被绿化，配置出奇趣的植被岛屿，抽象出"三江之岛"独有的自然风貌。

The landscape of Jiamusi Wanda Plaza represents the natural beauty of the scattered islands and golden river in "Sanjiang Plain (Three-River Plain)". The triangle shape stone paving at the mall main entrance reinforces the resplendent "Eastern Lights", creates a sense of arrival. The island planter is an impressive attraction in the Plaza. Local planting species has been carefully selected and placed here, showcasing an abstract landscape synonymous with "Sanjiang Island".

OUTDOOR PEDESTRIAN STREET
室外步行街

金街以"三江汇流"为主线，提取赫哲族特色的民俗文化作为亮点，如将弓箭抽象设计运用于灯具，铺装诠释的是鱼皮服饰的精美图腾。雕塑将赫哲人民生活场景进行抽象再现，尽显富饶的无限希望。在色彩方面借鉴了俄罗斯民族的风格，形成丰富的色彩变化，突出金街的丰富性、文化性和趣味性。

The design concept of Golden Street follows the theme of "Three-river Confluence", highlighted with "Hezhe Culture". The unique shapse of bow and fish skin pattern have been reinterpreted into the design of street lighting and paving. Sculptures in the plaza show the daily life scene of Hezhen people in an abstract way and represents the hope for prosperity. The Russian style color palette enriches the cultural characteristics of Golden Street.

D

WANDA HOTELS
万达酒店

WANDA COMMERCIAL PLANNING 2015

01 WANDA REIGN CHENGDU
成都万达瑞华酒店

时间：2015 / 12 / 18	**OPENED ON** : 18th DECEMBER, 2015
地点：四川 / 成都	**LOCATION** : CHENGDU, SICHUAN PROVINCE
建筑面积：4.16万平方米	**FLOOR AREA** : 41,600 M²

PROJECT OVERVIEW
酒店概况

成都万达瑞华酒店位于成都市中央商务核心区，紧临滨江中路与人民南路交叉口东北角，毗邻最繁华的盐市口、春熙路商圈，位置优越，交通便利。建筑总高度154米，包含奢华酒店（二十六至三十五层为酒店客房，三十六层为空中大堂）、甲级写字楼和定制商业三种业态。酒店大堂位于塔楼最顶层，层高9米，气势恢宏，可360度俯瞰周边城市景色。六层会所沿江设出挑平台，顶部配自动伸缩雨篷，在平台上锦江美景尽收眼底。

Ideally located in the city's CBD, close to the northeast corner at the intersection between Binjiang Mid Road and Renmin South Road, and adjacent to Yanshikou & Chunxi Road, Chengdu's main shopping area, Wanda Reign Chengdu enjoys an advantageous location and convenient transportation. The 154m high hotel accommodates three business types, including luxury hotel (hotel rooms are provided on the 26F-35F and sky lobby the 36F), grade-A office and customized commerce. With floor height of 9m, its spectacular lobby standing on the top floor of the tower can offer you all-round look over the surrounding city landscape. And its club on the fifth floor is provided with a cantilevered platform covered by an auto-retractable canopy, on which you may have a panoramic view of the splendid Jin River.

01 成都万达瑞华酒店总平面图
02 成都万达瑞华酒店外立面

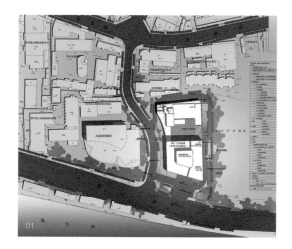

FACADE DESIGN
酒店外装

建筑外立面设计采用现代手法将传统地域文化完美演绎，以川西文化中的"竹"为基调，竖向构件的微妙变化配上简洁的建筑体量，使立面造型挺拔向上，宛如"节节高升"的竹子，冲入云霄，象征万达不断的发展和繁荣，成为该区域的标志性建筑。

By incorporating virtues of modern design techniques, the hotel facade perfectly interprets the traditional regional culture. Primarily based on "Bamboo" specific to the Western Sichuan culture, the facade presents a tall and straight shape via delicate variation of its vertical members and simple building massing, looking as if the unceasingly growing bamboo soaring to the sky and signifying continuous progress and prosperity of Wanda Group. The hotel is set to be a landmark of this region.

03-06 成都万达瑞华酒店实景照片

07 成都万达瑞华酒店入口
08 成都万达瑞华酒店包房内装

PART D · WANDA HOTELS 万达酒店 · 233

INTERIOR DESIGN
酒店内装

09

09-11 成都万达瑞华酒店屋顶花园
12 成都万达瑞华酒店水景

LANDSCAPE DESIGN
酒店景观

酒店景观围绕川西文脉，首层入口景观以"太极蜀韵"为主题的背景墙作为核心，配以汉白玉石等高档主材，辅以成都市市花《芙蓉》主雕，以强烈的特色形成首层入口迎接之势。面积达1200平方米的五层中式屋顶花园，以写意构图和多样化的园林手法，将小桥流水、亭台竹韵、湖石花木多种元素整合打造，步移景异，悠然雅致。

The hotel landscape centers on Western Sichuan context. Its entrance landscape on the ground floor highlights the background wall themed with "Sichuan Taiji", and employs high-grade materials such as white marble and main sculpture of "Lotus", the city flower of Chengdu, presenting a highly featured reception function. Covering an area of 1,200 square meters, the five-floor Chinese style roof garden utilizes freehand sketching and diversified gardening techniques to construct a series of elements (e.g. bridges, streams, pavilions, bamboos, lakeside rocks, flowers and woods) in an integrated manner, resulting in an "One Step, One Scene" effect and leisurely and graceful perception.

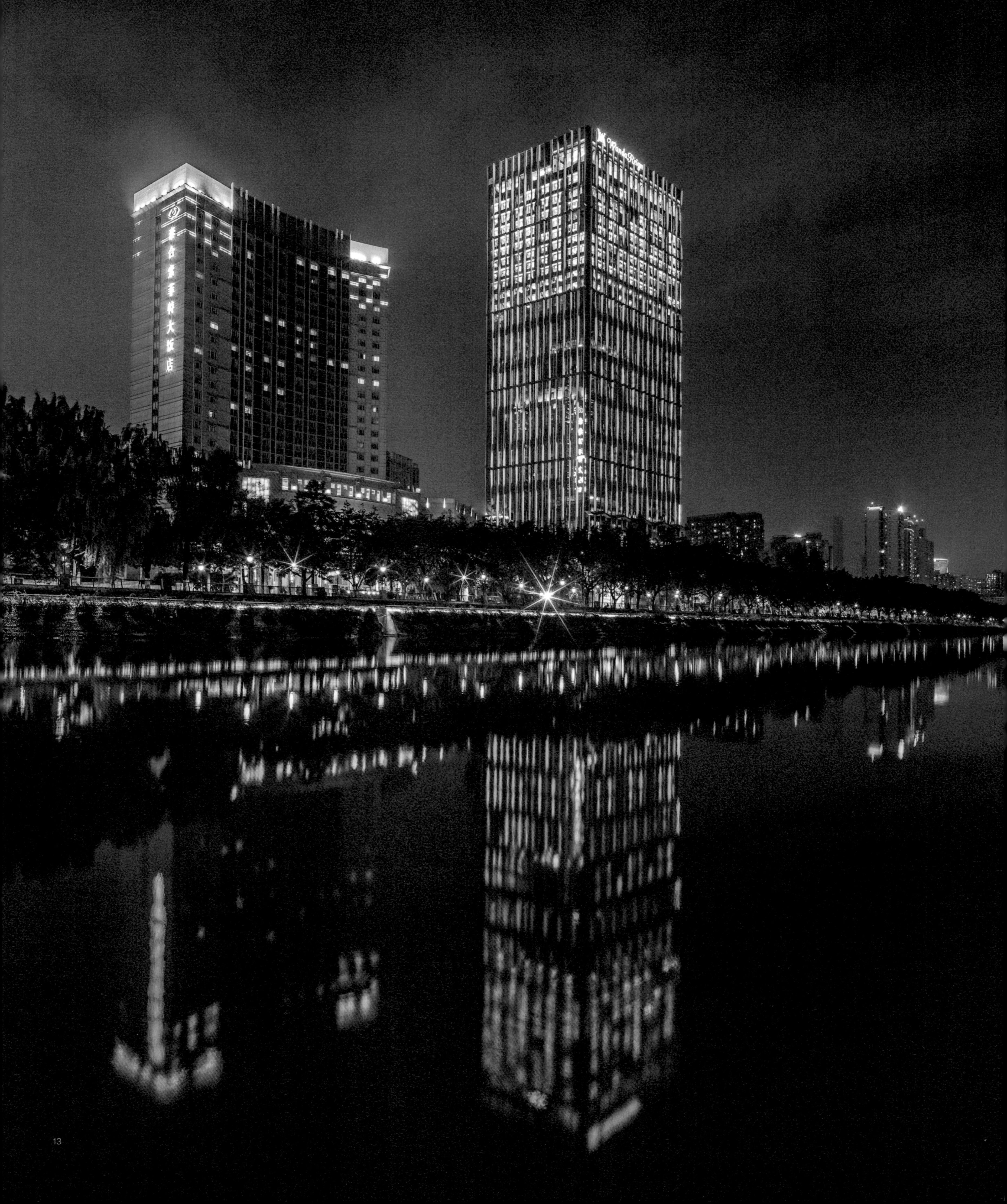

13

| PART D | WANDA HOTELS 万达酒店 | 237 |

13-14 成都万达瑞华酒店夜景

NIGHTSCAPE DESIGN
酒店夜景

02 WANDA VISTA HOHHOT
呼和浩特万达文华酒店

时间：2015 / 11 / 30
地点：内蒙古 / 呼和浩特
建筑面积：4.3 万平方米

OPENED ON : 30th NOVEMBER, 2015
LOCATION : HOHHOT, INNER MONGOLIA AUTONOMOUS REGION
FLOOR AREA : 43,000 M²

01 呼和浩特万达文华酒店总平面图
02 呼和浩特万达文华酒店立面图
03 呼和浩特万达文华酒店外立面

PROJECT OVERVIEW
酒店概况

呼和浩特万达文华酒店地处呼和浩特核心主干道新华大街，毗邻大型综合购物中心呼和浩特万达广场，交通极为便利，地理位置十分优越。酒店作为万达酒店及度假村旗下产品，在达到国际化设施和服务标准的同时，将塞外草原、蒙古包和传统蒙古族"祥纹"等民族元素完美贯穿于设计之中。酒店设有313间装修现代、设施齐全的豪华客房及套房。

Located in the Xinhua Avenue, the main trunk road of Hohhot, and adjacent to Hohhot Wanda Plaza (a large-scale complex shopping center), Wanda Vista Hohhot enjoys a convenient transportation and an advantageous geographical location. As a product of Wanda Hotels & Resorts boasting of international facility and service standard, the hotel hosts 313 well-equipped and luxurious guest rooms and suites with modern decoration, and also ingeniously incorporates such ethic elements as the Northern Frontier Plain, Mongolian yurt and Mongolian traditional "Auspicious Pattern" into its design.

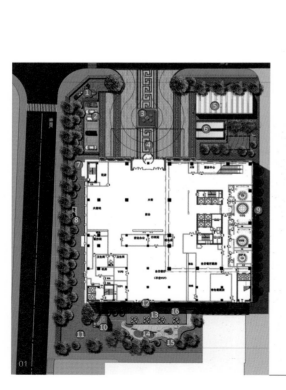

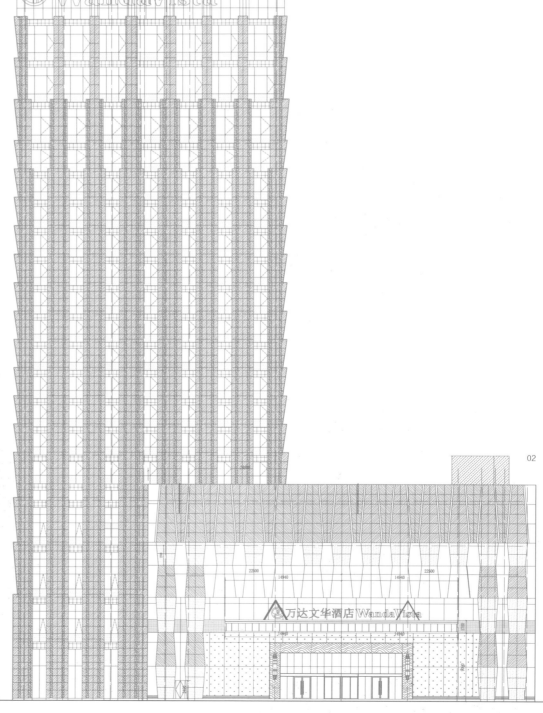

FACADE DESIGN
酒店外装

酒店塔楼立面和细节上借鉴了"蒙元文化"的元素，使酒店具有地方文化特色。塔楼层层外挑的造型充分体现中式传统木塔建筑的造型特点——"檐出深远，雄健有力"。裙楼立面采用钻石形石材与玻璃体穿插的造型，表现了文华酒店的奢华档次。正门大堂雨篷檐口装饰纹样采用蒙古族特有的"哈木尔云纹"，彰显酒店的地域特点及细节装饰上的考究。酒店成为运用当代设计手段和建筑材料重新诠释"蒙元文化"精髓的范例。

To endow the hotel with the local culture features, its tower facade and details are drawing on the elements of "Mongolian Culture in the Yuan Dynasty". The tower employs layered overhanging shape, fully reflecting the modeling characteristics of the Chinese traditional wooden towers-deeply overhung eave, robust and powerful. The podium facade applies diamond-shaped stones and glasses alternatively, demonstrating the luxury of the hotel. The canopy eave at the main entrance lobby is decorated with the "Mongolian Shamuer Pattern", highlighting regional characteristics and attentive decoration on details of the hotel. All these together make this hotel a paradigm of re-interpreting the essence of "Mongolian Culture in the Yuan Dynasty" with contemporary design methods and building materials.

04 呼和浩特万达文华酒店夜景
05 呼和浩特万达文华酒店外立面

06 呼和浩特万达文华酒店入口
07 呼和浩特万达文华酒店大堂
08 呼和浩特万达文华酒店总统套房门厅
09 呼和浩特万达文华酒店总统套房客厅

PART D　　WANDA HOTELS
万达酒店

INTERIOR DESIGN
酒店内装

10

11

10 呼和浩特万达文华酒店水景
11 呼和浩特万达文华酒店景观特写
12 呼和浩特万达文华酒店景观
13-14 呼和浩特万达文华酒店主入口喷泉

LANDSCAPE DESIGN
酒店景观

酒店景观以"蒙元文化"为线索，以"青城"为设计基调，以长调的韵律划分景观空间，营造独具"青城"特色的景观。酒店"前广场"带有浓郁蒙古文化的水景，从造型、色彩、构图和纹样上彰显文华酒店的高贵和地域特色；咖啡厅"外场"景观，以枯山水写意的设计手法彰显景观艺术美；酒店"后广场"的水幕墙是景观的经典收尾，将灵动的流水、幽静的水面与艺术沙石组合在一起，形成美妙的视觉画面，凸显文华酒店的内涵和典雅。

Threaded by the "Mongolian Culture in the Yuan Dynasty", designed based on "Blue City" (meaning of the name of the city in Mongolian), and divided with rhythm of Mongolian long tune, the hotel landscape is uniquely characterized by the theme of "Blue City". With its waterscape of rich Mongolian culture atmosphere, the "front square" of the hotel manifests dignity and regional characteristics of the hotel in terms of modeling, color, composition and pattern. Through the design method of freehand sketching of Japanese rock garden, the "outside" landscape of cafe takes on artistic beauty. In combination with running water, secluded water surface and artistic gravel, the water curtain wall of the hotel's "back square", the classic ending of the landscape, presents a fabulous visual image and highlights the cultural deposits and elegance of the hotel.

03 WANDA REALM LIUZHOU
柳州万达嘉华酒店

时间：2015 / 11 / 27　　**OPENED ON**：27th NOVEMBER, 2015
地点：广西 / 柳州　　　**LOCATION**：LIUZHOU, GUANGXI PROVINCE
建筑面积：3.75万平方米　**FLOOR AREA**：37,500 M²

01 柳州万达嘉华酒店鸟瞰图
02 柳州万达嘉华酒店总平面图
03 柳州万达嘉华酒外立面

PROJECT OVERVIEW
酒店概况

柳州万达嘉华酒店位于柳州城中万达广场，坐落于柳州市城中区文兴路与东环大道交叉口。酒店总建筑面积3.75万平方米，地上17层，地下2层。酒店共设有285间典雅的客房及套房，为商务人士创造温馨、惬意的居住体验。酒店拥有提供环球美食及外带早餐的"美食汇"全日餐厅、将精美粤菜与地方菜巧妙结合的"品珍"中餐厅以及"和"日式料理特色餐厅。1200平方米无柱大宴会厅，配以92平方米高清LED显示屏，为广大宾客提供国际一流水准的宴会服务。

With a gross floor area of 37,500 square meters, 17 floors above ground and 2 floors underground, Wanda Realm Hotel Liuzhou is located in Liuzhou Chengzhong Wanda Plaza that lies at the intersection between Wenxing Road, Chengzhong District, Liuzhou and East Ring Road. The hotel hosts 285 elegant guest rooms and suites, offering warm and cozy accommodation atmosphere for business travelers, and has all day "Café Vista" converging delicacy worldwide and serving take-away breakfast, "Zhen" Chinese Restaurant ingeniously gathering fine Cantonese cuisine and local cuisine together, and Japanese restaurant "Yamato". The hotel also provides a non-column grand banquet hall of 1,200 square meters, equipped with 92 square meters HD LED screen that can cater for banquets of internationally first-class standard.

04

FACADE DESIGN
酒店外装

酒店立面设计采用白色铝板作为主要材料，网格式单元框架为主要构成元素。通过统一的颜色和大小不一、错落有致的排列形成了均衡、稳定而又不失活泼的立面形体，良好的比例和尺度对单元框架加以有序阵列排列，突出其造型变化，体现了酒店挺拔的立面形象。

Mostly made of white aluminum plates, the facade of the hotel is largely constructed with the element of grid-shaped unit frame, whose well-proportioned arrangement in unified color yet varied sizes contributes to a balanced, stable yet lively facade image. Moreover, the modelling variation making prominent by an orderly layout in appropriate proportion and scale strengthens the towering image of hotel facade.

04 柳州万达嘉华酒店入口
05 柳州万达嘉华酒店大堂
06 柳州万达嘉华酒店电梯间

INTERIOR DESIGN
酒店内装

LANDSCAPE DESIGN
酒店景观

在酒店景观设计中，着重采用地方民族特色的元素，取得令人满意的效果。桂中壮锦——利用"壮锦"这种最能代表壮族民族手工艺的产品，体现当地的文化与艺术；奇石——酒店前广场以"奇石"结合水景，营造"仁者乐山，智者乐水"的意境；雕塑——观赏性与文化性融为一体，与酒店高贵的品质相得益彰；铺装——以编织中的壮锦形式呼应地方文化；竹子——利用当地特产竹子、竹编，结合柳州自然山水及溪流，营造静谧的酒店氛围。

In the landscape design of the hotel, a great emphasis is laid on the following elements bearing local ethic characteristics to harvest a desirable effect. Central Guangxi Zhuang brocade, the most representative ethic handicraft art product in the Zhuang ethnic group, is used to display the local culture and art. Odd-shaped stones in front of the hotel square, combined with waterscape, deliver an artistic conception of "the wise enjoy the waters, the benevolent enjoy the mountains". Sculptures fusing ornamental value and cultural meaning are in concord with dignified quality of the hotel. Pavement in the form of weaving Zhuang brocade echoes with local culture. Bamboo and its weaving, the local specialties, coupled with natural landscape and streams of Liuzhou, render a kind of tranquil hotel atmosphere.

07 柳州万达嘉华酒店水景
08 柳州万达嘉华酒店景观雕塑
09 柳州万达嘉华酒店夜景

NIGHTSCAPE DESIGN
酒店夜景

04 WANDA REALM TAI'AN
泰安万达嘉华酒店

时间：2015 / 08 / 21　**OPENED ON**: 21st AUGUST, 2015
地点：山东 / 泰安　**LOCATION**: TAIAN, SHANDONG PROVINCE
建筑面积：4万平方米　**FLOOR AREA**: 40,000 M²

OVERVIEW OF HOTEL
酒店概况

泰安万达嘉华酒店坐落于泰安市时代发展轴上，位于泰安万达广场西南角，地理位置优越，交通便利，是一座建筑高度150米的超高层建筑（一至四层、二十四至三十五层为酒店，六至二十二层为甲级写字楼）。酒店建筑规模4.0万平方米，达到五星级标准，包含300余套宽敞舒适的客房，以及全日餐厅、特色餐厅、中餐厅、宴会厅等大气典雅的就餐设施，同时具有游泳、健身、美容等休闲娱乐功能。

Situated in the Tai'an's Times development line and the southwest corner of Tai'an Wanda Plaza, the super high-rise Wanda Realm Tai'an with building height of 150m enjoys advantageous location and convenient transportation (hotel section is arranged on the 1F-4F and 24F-35F and grade-A office building on the 6F-22F). With a gross floor area size of 40,000 square meters, the five-star hotel is furnished with more than 300 spacious and comfortable guest rooms, grand and elegant catering facilities, such as an all-day restaurant, a Chinese restaurant, a specialty restaurant and a banquet hall, etc., as well as recreational functions including swimming, fitness and cosmetology.

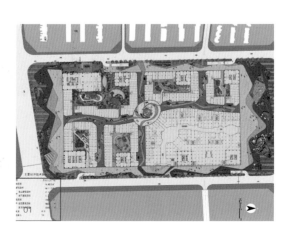

01 泰安万达嘉华酒店总平面图

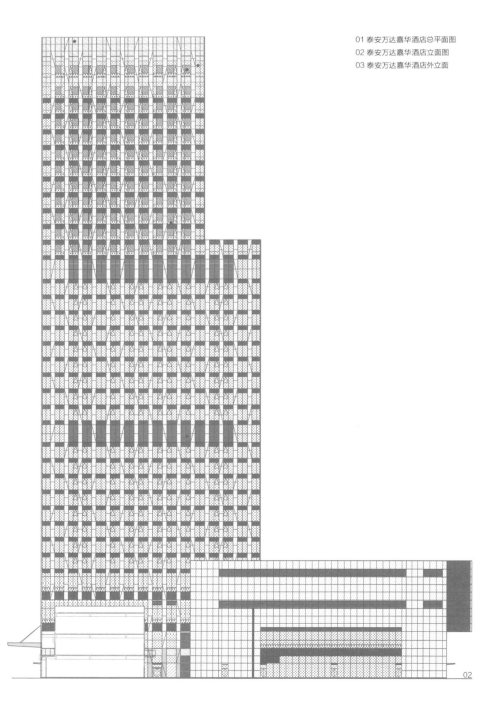

02 泰安万达嘉华酒店立面图
03 泰安万达嘉华酒店外立面

FACADE DESIGN
酒店外装

酒店立面采用现代设计元素，通过铝板和玻璃两种材料的搭配，配合铝板独特的折线造型，将现代城市的纯净、简洁、明朗的风格体现出来。铝板幕墙采用独特的折线造型，象征着泰山的峰峦叠嶂、也象征泰山的不屈性格，将地域文化特性融入现代建筑之中。玻璃幕墙采用高反射率的中空镀膜玻璃，在满足节能环保的同时，满足内部空间的私密性。

Incorporating modern design elements, the hotel facade design attempts to render the pure, concise and cheerful style of modern cities through the collocation of two kinds of materials, aluminum plates and glasses, coupled with unique fold line modeled of the former material. Aluminum plate curtain wall featuring fold lines signifies hills and cliffs and unyielding character of the Mount Tai, an example of integrating regional cultural features into the modern architecture. Whereas the glass curtain wall uses hollow coated glasses with high reflectivity, which are both energy efficient and environmental friendly and meanwhile, securing interior privacy.

PART **D** | WANDA HOTELS 万达酒店 | 255

INTERIOR DESIGN
酒店内装

04 泰安万达嘉华酒店外立面
05 泰安万达嘉华酒店入口
06 泰安万达嘉华酒店全日餐厅
07 泰安万达嘉华酒店大堂

08 泰安万达嘉华酒店入口
09 泰安万达嘉华酒店水景
10 泰安万达嘉华酒店喷泉
11 泰安万达嘉华酒店夜景

LANDSCAPE DESIGN
酒店景观

景观设计注重细节刻画。入口水景用起伏的折线勾勒连绵山峰的轮廓线；水台上雕刻了云形图案；玻璃材质的运用，让光影和虚实感更显写意。酒店后院的水景立面上的景石和石桩，营造苍松、巨石屹立于云间的意境。北侧的景墙运用"西王母驾车"等文化题材的雕塑，采用"枯山水"的格局，显得安静典雅不失底蕴。

The landscape design infuses painstaking efforts to details. The waterscape at the entrance outlines stretching peaks with rolling fold line; the water platform is carved with cloud patterns; the application of glass material generates a more joyful feeling of light & shadow and the real & virtual. The waterscape at backyard applies scenic stones and stone piles on its facade, building the atmosphere of green pines and huge rocks erecting among clouds. The landscape wall on the north employs sculptures with cultural themes like Queen Mother of the West (figure in Chinese mythology)'s Drive, laid out in the form of "Japanese rock garden", seeming to be peaceful, elegant yet containing cultural meanings.

NIGHTSCAPE DESIGN
酒店夜景

05 WANDA REALM HUANGSHI
黄石万达嘉华酒店

时间：2015 / 07 / 03　　**OPENED ON**：3rd JULY, 2015
地点：湖北 / 黄石　　　**LOCATION**：HUANGSHI, HUBEI PROVINCE
建筑面积：3.25 万平方米　**FLOOR AREA**：32,500 M²

01 黄石万达嘉华酒店鸟瞰图
02 黄石万达嘉华酒店总平面图
03 黄石万达嘉华酒店立面图

PROJECT OVERVIEW
酒店概况

黄石万达嘉华酒店位于黄石市花湖大道，紧邻城市中轴线湖滨路，地理位置优越，交通便利。总建筑面积3.25万平方米，总高度70米，地下2层，地上15层。酒店共有客房263间，恒温泳池1座，多功能会议室5间，健身房1间，瑜伽房1间，特色餐散座厅1间、特色餐包间3间；中餐散座厅1间、中餐包间6间。

Ideally located in Huahu Avenue, Huangshi City and next to Hubin Road, the city central axis, Wanda Realm Huangshi occupies an advantageous location and convenient transportation. With a gross floor area and total height of 32,500 square meters and 70m respectively, the hotel has two floors underground and 15 floors above ground, and consists of 263 guest rooms, one heated swimming pool, five multi-functional meeting rooms, one fitness room, one yoga room, a specialty restaurant with one extra seats and three private rooms, and a Chinese restaurant with one extra seats and six private rooms.

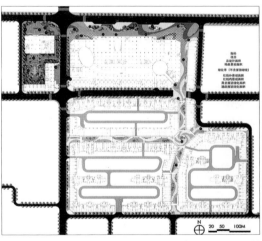

02

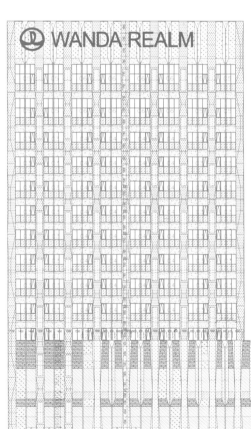

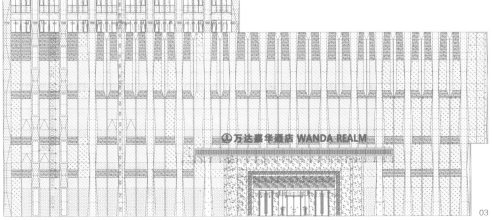

03

04 黄石万达嘉华酒店入口
05 黄石万达嘉华酒店大堂

FACADE DESIGN
酒店外装

黄石古称"大冶",取"大兴炉冶"之意。黄石境内矿产丰富,冶炼业发达,素有"中国观赏石之城"、"青铜故里"之称。当地盛产的晶体,呈规则有机排列,具有强烈的序列感,可以形成强烈的视觉冲击。立面设计灵感"日精月华"便来自于这些晶体矿石,以晶体形态作为立面构成元素,采用多种切割方法,使立面产生多变的光影效果。

Huangshi, called "Daye" in the past meaning "large-scale of smelting", is cradle to mineral resources and developed smelting industry, giving its name of "City of Decoration Stones in China" and "Home of Bronze". The local abundant crystals arranged in regular and organic manner present a strong sense of sequence and massive visual impact accordingly. Inspired by these crystal ores, the facade design themed with "the Sun and Moon's Essence" takes crystal shape as its constituent element and employs a variety of cutting methods, so that changeable light & shade effect is generated.

INTERIOR DESIGN
酒店内装

06 WANDA REALM ANYANG
安阳万达嘉华酒店

时间：2015 / 07 / 25　　**OPENED ON** : 25th JULY, 2015
地点：河南 / 安阳　　　**CITY** : ANYANG, HENAN PROVINCE
建筑面积：3.25万平方米　**FLOOR AREA** : 32,500 M²

01 安阳万达嘉华酒店外立面
02 安阳万达嘉华酒店入口
03 安阳万达嘉华酒店总平面图

PROJECT OVERVIEW
酒店概况

安阳万达嘉华酒店位于安阳市中华路南段29号,总建筑面积3.25万平方米,地上16层,地下2层。酒店共设有286间典雅的客房及套房,为商务人士创造温馨、惬意的居停感受。酒店拥有提供环球美食及外带早餐的"美食汇"全日餐厅、将精美粤菜与地方菜巧妙结合的"品珍"中餐厅以及以川菜为主的"辣道"特色餐厅;1000平方米无柱大宴会厅为市内最大,可提供国际一流水准的宴会服务。

Located in No.29 Zhonghua Road South Wenfeng District, Anyang, Wanda Realm Anyang has a gross floor area of 32,500 square meters, 16 floors above ground and 2 floors underground. The hotel hosts 286 elegant guest rooms and suites offering warm and cozy accommodation atmosphere for business travelers, and has all day "Café Vista" converging delicacy worldwide and serving take-away breakfast, "Zhen" Chinese Restaurant ingeniously gathering fine Cantonese cuisine and local cuisine together, and Chili & Pepper specialty restaurant. The hotel also provides a 1,200 square meters non-column grand banquet hall that can cater for banquets of internationally first-class standard.

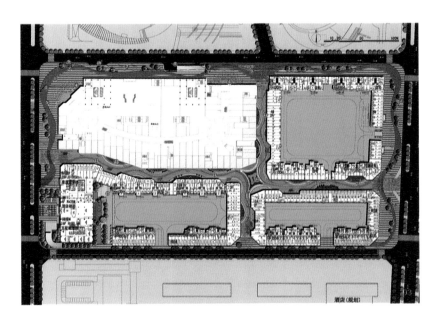

FACADE DESIGN
酒店外装

酒店在外立面设计上以"甲骨文化"作背景，汲取三千年文化古都的精华，采用富有传统文化含义的刀币形象作为母题，特点鲜明。刀币母题根据酒店客房错落排列，并辅以现代大气设计手法和铝板、玻璃等现代感强烈的材料，达到了形式和功能、手法和材料、传统地域文化和现代建筑的完美统一，成为古老与现代融合的典范。

With a "Chinese Oracle" setting, the featured facade design of the hotel draws on the essence of the three-thousand-year-old ancient cultural capital and takes the knife money image full of traditional culture meaning as its theme. The knife money being disorderly laid as per the hotel guest rooms, supplemented with grand modern design method, as well as materials of strong modern sense like aluminum plates and glasses, the hotel facade attains ideal unity in form and function, method and material, and traditional regional culture and modern building, serving as an example of integrating the ancient and modern times.

04 安阳万达嘉华酒店外立面
05 安阳万达嘉华酒店大堂
06 安阳万达嘉华酒店景观

04

INTERIOR DESIGN
酒店内装

LANDSCAPE DESIGN
酒店景观

酒店景观设计以"洹河印记"为概念设计主题,将洹河岸边的文化及典故加以提炼。酒店前场灵动的水景寓意安阳"母亲河"的生机、活力;铺装上"云雷纹"图案则紧扣殷商青铜文化,提取青铜纹路,象征了一条探寻玉石之路。植物群落因地制宜、塑造空间个性。酒店后场点缀着浮雕景墙,运用秦前文字"甲骨文",景墙浮雕主题"十二生肖字形演变图",突出秦前时代至现代的字形演变过程。

Following the theme of "Huan River Mark", the landscape design of the hotel extracts culture and allusions along the Huan River. In front of house, the dynamic waterscape implies vitality and vigor of the "Mother River" of Anyang; "Cloudscape" on pavement drawing on bronze pattern from the Bronze culture of the Shang Dynasty signifies the path of exploring jades; plant community adapting to circumstances moulds character of the space. In back of house, interspersed with relief that follows the theme of "Glyph Evolution Figure of the Chinese Zodiac", the landscape wall adopts the "Oracle", the pre-Qin characters and highlights the glyph evolution from the Pre-Qin period to the modern times.

07 WANDA REALM GUANGYUAN
广元万达嘉华酒店

时间：2015 / 06 / 05　　**OPENED ON**：5th JUNE, 2015
地点：四川 / 广元　　　**LOCATION**：GUANGYUAN, SICHUAN PROVINCE
建筑面积：3.4万平方米　**FLOOR AREA**：34,000 M²

01 广元万达嘉华酒店鸟瞰图
02 广元万达嘉华酒店总平面图
03 广元万达嘉华酒店外立面

PROJECT OVERVIEW
酒店概况

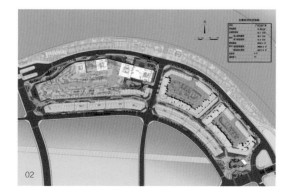

广元万达嘉华酒店坐落于广元市核心商圈万源新区的万达广场，总建筑面积3.4万平方米，地上17层，地下2层，拥有285间宽敞舒适的客房及套房，氛围高雅、配套豪华。"美食汇"为国际风味美食及自助餐厅；"品珍"中餐厅以华美的氛围、精致的美味佳肴而闻名；"24小时精选"呈献"家一般的温暖"。此外，酒店游泳池、健身中心、商务中心、会议室、宴会厅、停车场等高标准设施一应俱全。

Located in the Wanda Plaza that is constructed on Wanyuan New District, the core business district of Guangyuan City, Wanda Realm Guangyuan has a gross floor area of 34,000 square meters, 17 floors above ground and 2 floors underground. The hotel hosts 285 spacious and comfortable guest rooms and suites rendering elegant atmosphere and luxurious supporting facilities, and has "Café Vista", a buffet restaurant converging delicacy worldwide, "Zhen" Chinese Restaurant known for its splendid environment and exquisite delicacies, and Wanda Realm's unique "24-hours Selected" bringing you warmth of home. The hotel also boasts of its completed high-grade facilities, an encompassing swimming pool, a fitness center, a business center, meeting rooms, a banquet hall, and its parking lot, etc.

FACADE DESIGN
酒店外装

在现代大气的基础上融合"蜀道文化",辅以"石窟文化"的构成做一些背景的演变和应用。在细节上,提取广元"三国文化",以"水"为题,用"波"型曲线的元素贯穿始终,强调"竖向线条"宽窄变化来塑造立面造型,形成大尺度的流畅曲线,打造标志性视觉效果。穿孔板的运用是在玻璃与铝板"虚"、"实"之间增加一个半透明的设计元素,结合其"波浪"造型,为整个建筑增添现代气氛。

In background, incorporation of the "Sichuan Road Culture" into the grand modern style, complemented by "grotto culture", brings evolution and application. In details, sourcing idea from the "culture of Three Kingdoms", the facade follows the theme of "water" and applies "wave" curves throughout. The design strives to mould facade shape mainly through width change of "vertical lines", so that smooth curves of large scale are taken shape to render iconic visual effect. The adding of a translucent element, perforated plates, between the virtual (glasses) and the real (aluminum plates), together with its "wave" shape, strengthens the modern sense of the whole building.

INTERIOR DESIGN
酒店内装

LANDSCAPE OF HOTEL
酒店景观

酒店景观延续建筑的风格，注重细节设计，运用木板、水面、涌泉、艺术廊架，营造闹中取静的休闲空间，提升酒店文化品位；同时将地方文化元素以简练、现代的方式融入设计当中，力求在促进商业氛围的同时宣传本土文化。

Inheriting the detail-oriented building style, the hotel landscape builds a tranquil recreation space via planks, waters, fountains and art gallery frames for an improved cultural taste of the hotel. The landscape design, meanwhile, incorporates local culture elements in a concise and modern way, to promote commercial atmosphere and publicize local culture.

04 广元万达嘉华酒店入口
05-06 广元万达嘉华酒店外立面特写
07 广元万达嘉华酒店大堂
08 广元万达嘉华酒店喷泉

08 WANDA REALM NEIJIANG
内江万达嘉华酒店

时间：2015 / 06 / 26　　OPENED ON：26th JUNE, 2015
地点：四川 / 内江　　　LOCATION：NEIJIANG, SICHUAN PROVINCE
建筑面积：2.8万平方米　FLOOR AREA：28,000 M²

01 内江万达嘉华酒店外立面
02 内江万达嘉华酒店总平面图
03 内江万达嘉华酒店外立面

PROJECT OVERVIEW
酒店概况

内江万达嘉华酒店位于内江市汉安大道，毗邻万达广场和国际会展中心。酒店总建筑面积3.26万平方米，地上16层，地下2层；酒店拥有263间客房和套房，设有面积为1100平方米、层高10米的无柱式大宴会厅，配备了70平方米高清LED显示屏，以及4个配备最先进科技和视听设备的多功能厅。

Located in Han'an Road, Neijiang City, Wanda Realm Neijiang is next to Wanda Plaza and International Convention Center. With gross floor area of 32,600 square meters, the hotel has 16 floors aboveground and 2 floors underground. It accommodates 263 guest rooms and suites, and provides a 10m high 1,100 square meters non-column banquet hall equipped with 70 square meters HD LED screen, and four multi-functional halls furnished with the state-of-the-art audio-visual equipment.

FACADE DESIGN
酒店外装

立面设计灵感来源于这座山水城市本身的特质,名为"山水云间"。主体建筑材料采用银灰色铝板,以铝板冷峻的色调将主体建筑表达为挺拔威严的山体,同时运用了蓝色和白色两种不同颜色不同透光率的玻璃来象征湛蓝的流水与律动的白云。光线折射、色彩和尺度的大胆运用,都为这一重要城市中心公共空间带来冲击与活力,创造合适的环境背景,"水绕云间"的画面宛如人间仙境。

Inspired by the very nature of the landscape city, facade design of the hotel is defined as "Landscape in Clouds". The main building adopts aluminum plates in cool color of silver grey to depict itself as a towering and solemn mountain, and blue and white glasses with different transmittance to signify clear blue waters and dynamic clouds. Benefited from such refraction and bold application of color and scale, this important downtown public space is endowed with vitality and impact, as well as a desirable environmental setting. The picture of landscape in clouds just looks like a fairyland.

INTERIOR DESIGN
酒店内装

04 内江万达嘉华酒店入口
05 内江万达嘉华酒店大堂
06-08 内江万达嘉华酒店景观

LANDSCAPE DESIGN
酒店景观

内江是著名国画大师张大千的故乡，以"大千故里，书画之乡"美誉全国。酒店景观设计以大千画作中的"山"、"水"作为视觉元素，刻画"水墨内江"。在酒店景观设计中，无论是铺装、水景还是小品，都凸显"山韵水意"之理念，强调与环境的融合，通过不同材料之对比（自然石、铜板、印刷玻璃），将设计融入"大千的山水画卷"之中。景观结合酒店功能空间，如前场落客区、后场全日餐外延、花园散步区等，同时植物设计以四川当地植物为主，具有蜀川特色。

Neijiang is the hometown of Chang Dai-Chien, a best-known Chinese painting master, and renowned nationwide for its reputation of "the home of Chang Dai-Chien calligraphy and painting". Visually displaying "Mountain" and "Water" found in works of Mr. Chang, landscape design of the hotel tries to depict an "Ink Neijiang". To this end, pavement, waterscape and landscape articles all highlight the idea of "charm of mountain and water", and the infusion with the environment. Through contrast of diverse materials (natural stones, copper plates, printed glasses), the design is integrated with "landscape scroll of Chang Dai-Chien". Meanwhile, the landscape is connected to functional space, such as drop-off area of front of house, extension of all day dining restaurant and pedestrian space of garden. The use of local plants adds Shichuan characteristics to the landscape.

09 WANDA REALM DONGYING
东营万达嘉华酒店

时间：2015 / 08 / 17　　**OPENED ON**：17th AUGUST, 2015
地点：山东 / 东营　　　**LOCATION**：DONGYING, SHANDONG PROVINCE
建筑面积：3.78 万平方米　**FLOOR AREA**：37,800 M²

01 东营万达嘉华酒店立面图
02 东营万达嘉华酒店外立面
03 东营万达嘉华酒店总平面图

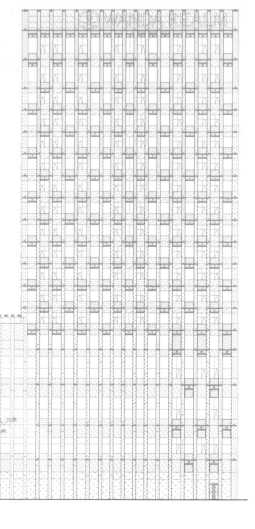

01

PROJECT OVERVIEW
酒店概况

东营万达嘉华酒店坐落于东营市东营区，交通极为便利，为标准五星级酒店。酒店总建筑面积3.78万平方米，其中地下2层、裙楼4层、塔楼13层，共286间客房。酒店拥有提供环球美食及外带早餐的"美食汇"全日餐厅，以粤菜为主、辅以本地特色美食的"品珍"中餐厅以及以川湘辣味为主的"辣道"特色餐厅。市内最大的1200平方米无柱大宴会厅，配以80平方米高清LED显示屏，为广大宾客提供国际一流水准的宴会服务。

Located in Dongying District, Dongying City, Wanda Realm Dongying, a standard five-star hotel, enjoys a convenient transportation. With a gross floor area of 37,800 square meters, the hotel has two underground floors, a 4-floor podium and a 13-floor tower, accommodating totally 286 guest rooms. There is all-day "Café Vista" converging delicacy worldwide and serving take-away breakfast, "Zhen" Chinese Restaurant ingeniously gathering fine Cantonese cuisine and local cuisine together, and Chili & Pepper specialty restaurant serving Sichuan and Hunan spicy cuisine. The hotel also provides a non-column grand banquet hall of 1,200 square meters, equipped with a 80 square-meter HD LED screen that can cater for banquets of internationally first-class standard.

FACADE DESIGN
酒店外装

酒店外立面构思来源于水波纹。酒店外立面以曲折的竖向铝板装饰线条为主，辅以跳色布置的窗槛墙，形成一种错落有致、富有动感的造型。客房塔楼的竖向线条一直延续到裙房上，整个建筑外观效果显得非常统一和谐。入口位置通过门框式的造型强调出来，配合造型活泼的雨篷，更加强化了酒店主入口的效果。

Sourcing idea from water ripple, the hotel facade mainly applies zigzag aluminum panels decorative moulding in vertical direction, supplemented by spandrels in alternated colors, forming a well-proportioned and dynamic shape. Podium inherits the vertical lines of guest room tower, achieving a very harmonious appearance. Expressed in doorframe shape and combined with lively-shaped canopy, the main entrance of the hotel becomes conspicuous.

INTERIOR DESIGN
酒店内装

04 东营万达嘉华酒店入口
05 东营万达嘉华酒店大堂
06-07 东营万达嘉华酒店水景
08 东营万达嘉华酒店景观

LANDSCAPE DESIGN
酒店景观

酒店景观设计以"长河落日"为主题，整体简洁大气。酒店区域主入口设置对称主景大树强调酒店的门户仪式感。酒店入口以对景的形式设计跌水景观，气势如黄河奔腾。酒店后场全日餐区域通过木平台及景墙水景的设计形成围合空间，呈现五星酒店的品质与气质。主景叠水墙以简洁的手法体现酒店庭院的宁静与现代感：水幕跌落在开阔的镜面水景中，形成"黄河入海，长河落日"的意境。

The landscape design of the hotel takes the theme of "Sunset Meeting the River", simple and grand. The main entrance provides symmetrically arranged accent trees to emphasize its ceremonial sense of portal, and cascading landscape in opposite scenery to present the momentum of rushing Yellow River. All-Day-Dining area at back of house is a space enclosed by wooden platform and landscape wall waterscape, reflecting the quality and disposition of the five-star hotel. The accent cascading wall designed in a simple way to display the tranquil and modern sense of courtyard, as water screen falling on vast mirror-like waterscape presents a vision of "Yellow River runs into the Sea and Sunset meets the River".

06

07

08

10 WANDA REALM FUYANG
阜阳万达嘉华酒店

时间：2015 / 08 / 29　　**OPENED ON**：29th AUGUST, 2015
地点：安徽 / 阜阳　　　　**LOCATION**：FUYANG, ANHUI PROVINCE
建筑面积：3.6万平方米　　**FLOOR AREA**：36,000 M²

PROJECT OVERVIEW
酒店概况

阜阳颍州万达嘉华酒店坐落于阜阳市中心的颍州区，交通便利，为标准五星级酒店。酒店总建筑面积3.6万平方米，其中地下2层、裙楼4层、塔楼13层，共286间客房。"品珍"中餐厅以精美粤菜和本地特色菜为主，"辣道"特色餐厅出品川湘菜特色美食，其他服务、娱乐设施也一应俱全。

Located in Yingzhou District, the center of Fuyang City, Wanda Realm Fuyang, a standard five-star hotel, enjoys a convenient transportation. With a gross floor area of 36,000 square meters, the hotel has two underground floors, a 4-floor podium and 13-floor tower, accommodating totally 286 guest rooms. There are "Zhen" Chinese Restaurant devoted to Cantonese flavors and local delicacies, Chili & Pepper specialty restaurant serving Sichuan and Hunan spicy cuisine, and other services and recreational facilities.

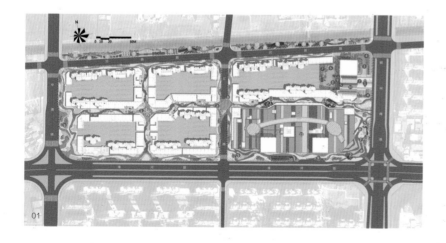

01

01 阜阳万达嘉华酒店总平面图
02 阜阳万达嘉华酒店外立面

03

03 阜阳万达嘉华酒店外立面
04 阜阳万达嘉华酒店入口
05 阜阳万达嘉华酒店夜景

FACADE DESIGN
酒店外装

建筑设计贯穿具有思想深度的设计哲学，在"传承——继承传统元素"、"联系——与大商业的联系"、"独特——独特的标志性建筑"和"未来——与阜阳共同走向未来"等四个方面加以思考和突破。设计将阜南闻名于世的"柳编"形象转化为结构形式之美加以表达。整体造型运用编织的形象手法，由上自下渐变，疏密有致，创造出丰富而有层次的立面造型。

The hotel design focuses on an insightful philosophy, making contemplation and breakthrough on four aspects: inheritance-inheritance of traditional elements, link-link to large commercial area, uniqueness-unique landmark building and future-A shared future with Fuyang. The design interprets the renowned Funan Willow image in the "beauty of structural form". The overall facade shape is weaved to have top-down gradation and well-arranged layout, making it abundant and distinctive in gradation.

NIGHTSCAPE DESIGN
酒店夜景

E
DESIGN AND CONTROL
设计与管控

WANDA COMMERCIAL PLANNING 2015

R&D AND APPLICATION OF "WANDA PLAZA STANDARDIZED DESIGN"
"万达广场标准化设计"的研发及应用

万达商业地产高级总裁助理兼商业规划研究院院长　叶宇峰

万达集团进行的"万达广场标准化设计"研发工作，不仅是集团展望2020年商业地产规模化高速发展的需要，更是降低商业地产开发成本、高效发展的需要。

作为国内最大商业地产开发企业的万达集团，其开发的万达广场，已经是国内商业广场的成功代表；但在目前商业地产的现状下，如何利用"标准化设计"，使万达广场保持在市场上的领先优势，进一步拓展市场，加强万达广场建设的竞争优势，提高万达广场的投资回报率，是万达集团一直在认真思考和积极研发的课题。

Wanda Group's R&D work on Wanda Plaza Standardized Design (hereinafter referred to as the "standardized design") is essential to both the Group's vision on rapid scale development of Commercial Properties in 2020 and Commercial Properties' call for reduced development cost and efficient progress.

Wanda Plaza developed by Wanda Group, China's largest estate enterprise in commercial properties, has set up a model of successful domestic commercial plazas. Yet under current situation of commercial properties, the issue lies in how to utilize the "standardized design" to maintain leading market edge, further expand market, and enhance competitive advantages and ROI of Wanda Plazas. It is also the very issue that Wanda Group has been deliberately pondering on and vigorously delving into.

一、"万达广场标准化设计"研发进程与发展

I. R&D PROCESS AND DEVELOPMENT OF STANDARDIZED DESIGN

"万达广场标准化"的研发工作早在2012年《万达限额设计建造标准（2012版）》就已经开始，并日臻完善，到2015年标准化程度已达到40%。"万达广场标准化"的研发，主要包括五方面的内容：万达建造标准（限额设计）、通用标准施工图集、标准单元模块、专项标准模块化、效果类标准设计样板库（表1）。

R&D work on Standardized Design can be traced back to 2012, during when *Construction Cost Standard of Wanda Quota Design (2012)* was issued. Thereafter, it has been constantly improved and in 2015, standardization degree reached 40%. R&D work on Standardized Design mainly consists of the following five aspects: Wanda construction standard (quota design), general standard construction atlas, standard unit module, special standard modularization and effect standard design prototype library (Table 1).

二、"万达广场标准化设计"研发成果及应用

II. R&D RESULTS AND APPLICATION OF STANDARDIZED DESIGN

1.万达建造标准（限额设计）
万达的建造成本管理紧紧围绕建造标准而展开——

1. WANDA'S CONSTRUCTION COST STANDARD (QUOTA DESIGN)

序号	标准化事项	标准化成果	完成时间	阶段标准化率	累计标准化率
1	建造标准（限额设计）	《万达广场定额设计技术标准》(2012版)	2012	5%	5%
2	通用标准施工图集	《通用标准施工图集》	2015-I	10%	15%
3	标准单元模块	采光顶、制冷机房、慧云机房等标准单元模块	2015-II	5%	20%
4	专项标准模块化	立面、内装、景观等效果类标准方案	2015-III	10%	30%
5	效果类标准样板库	效果类各级建造标准设计封样的样板库及电子库	2015—2016	10%	40%

(表1) "万达广场标准化"成果一览表

先按照万达建造标准，由多部门共同制定合理可行的目标成本，然后在项目开发过程中通过合约规划，对成本进行层层分解、落实；同时依据各项制度、过程管控措施保证不超目标成本，最终保证企业预期利润的实现，达到主动和事前控制的目的。

基于建造标准在成本控制中的核心作用，集团历来高度重视建造标准的制定，采取每两年修订一次的原则。从2011年开始颁布《2011版建造标准》，在2013年修订成《2013版建造标准》；2015年根据集团进一步控制成本的方针，在广泛调研实际开业、建设项目和市场行情，结合2015年5月1日正式实施的新《建筑设计防火规范》，兼顾直投模式，由规划院、集团成本部牵头集团其他各相关部门在《2013版建造标准》基础上，共同修编完成了《2015版建造标准》。本次修编工作有以下几个主要特点。

（1）标准细分原则——随着万达业务快速发展，万达广场开发范围从一线、二线拓展到三线城市，相应的建造标准由原来的四级调整为六级（即A+版、A版、A-版、B版、B-版和C版），可适应集团对于各级城市项目投资的要求，满足各级城市项目投资回报率的需要。

（2）结合设计标准化、模块化、产业化原则——为全方位优化设计施工成本，真正通过设计创造效益，本次修编工作结合设计科研，从标准化、模块化、产业化方面对项目进行研究。特别是对于新版《建筑设计防火规范》等涉及消防安全规范的事宜进行了重点修编，使新编建造标准既满足集团对于建造成本方面的控制要求，又满足新规范对于消防方面的要求，最终取得了满意的结果。

2. 通用标准施工图集

《直投万达广场土建施工图标准图集》分两阶段编制完成，第一阶段成果于2015年7月8日正式发布实施，第二阶段成果于2015年10月15日正式发布实施。《直投万达广场土建施工图标准图集》包括建筑、结构、暖通、给水排水、强电、弱电智能化这六个专业，对施工图设计进行标准化，以减少后续项目的工作量、缩短设计周期，保证设计质量，降低成本、保证项目整体开发工作的顺畅推进。

3. 标准单元模块

万达广场标准模块单元是在直投万达广场标准平面图的基础上提炼出来的标准模块图。该套模块图用于指导集团各部门及各设计单位快速选用标准模块进行设计，从而提高直投万达广场的标准化率，实现集团"BIM标准化设计"的战略目标。

The construction costs management of Wanda focuses closely on the Construction Cost Standard. Reasonable and workable target costs shall firstly be made by several departments in accordance with Wanda's Construction Cost Standard; then, in the process of project development, the costs will be broken down and put into practice step by step based on contract planning; meanwhile, it shall be ensured that the actual expenditures shall not exceed the target costs through the implementation of different regulations and process management and control measures; at last, the expected profit shall be achieved, which shall satisfy the demand of initiative and feed-forward control.

Considering the core function of construction cost standard in cost control, Wanda Group has always attached great importance to the standard, which is to be revised biennially. In 2011, *Construction Cost Standard 2011* was issued. In 2013, the revised edition *Construction Cost Standard 2013* was completed. In 2015, based on Group's guideline to further control cost and extensive research on projects actually opened, projects under construction and market condition, and combined with *Code for Fire Protection Design* officially implemented on May 1, 2015 and direct investment model, the compilation of *Construction Cost Standard 2015*, led by the Planning Institute and the Group's Cost Control Department and with the cooperation of other relevant departments, has been completed based on *Construction Standard 2013*. The said compilation has main characteristics as below:

(1) Principle of standardization segmentation: along with rapid development of business, Wanda Plazas also make a presence in the third-tier cities in addition to first-tier and second-tier ones. Accordingly, the original four-level construction cost standard is scaled up to six levels (i.e. Edition A+, Edition A, Edition A-, Edition B, Edition B- and Edition C), so as to satisfy the project investment and the ROI required by cities of different tiers.

(2) Combining principle of design standardization, modularization and industrialization: to optimize the costs of design & construction at all dimensions and to truly generate revenue by design, design research has been integrated into this compilation work, and the project has been studied from the perspective of standardization, modularization and industrialization. To make the newly-compiled Construction Cost Standard satisfy both the Group's requirements on construction costs control and the fire protection requirements of the new code, special and careful attention has been given to the revision of matters concerning fire safety specifications set out in the new edition of *Code for Fire Protection Design*, and finally achieves a satisfactory result.

2. GENERAL STANDARD CONSTRUCTION DRAWING ATLAS

Civil Construction Drawings Standard Atlas of Directly Invested Wanda Plazas are prepared in two phases. Results of the first phase were released and implemented on July 8, 2015 and the second phase on October 15, 2015. The atlas carries out standardization for construction drawings design from six specialties (architecture, structure, HVAC, plumbing, HV and ELV), aiming to reduce subsequent workload and design cycle, and ensure design quality, lower cost and smooth advance of whole project development.

Covering plans, sections and main nodes details in projects, the atlas can handle 80% of conventional construction and thus greatly shortens design cycle.

3. STANDARD UNIT MODULE

Wanda Plazas' standard unit modules refer to standard module drawings based on direct invested Wanda Plazas'

本套图集收录的是直投万达广场标准图中使用率高，且可以进行模块化设计的内容（如楼电梯、卫生间、管井等），方便各部门直接套用；它的适用性强，无论全标、类标、非标项目，均可以选择使用。

标准模块单元供直投万达广场在方案设计阶段选用，初步设计及施工图设计阶段需要根据项目实际情况进行调整后使用。图纸共包括三大类，分别是建筑篇、机电篇、景观篇。

如采光顶模块，万达在万达广场采光顶标准化设计基础上，研发了适应工厂化生产和装配化施工的工艺流程，将采光顶划分为七大系统，包括：围护系统、受力系统、消防系统、遮阳系统、照明系统、吊挂系统、防雷系统。对每个系统的每个部件进行标准化设计，最终形成《万达广场采光顶产业化标准设计图集》。对采光顶的围护结构、受力系统、消防系统等三方面进行了系统研究，之后通过工业化生产、装配式安装、信息化管理实现采光顶的标准化设计和模块化施工。目前，采光顶已完成产业化试点并取得突出效果。

4. 专项标准模块化

2015年完成了万达广场的效果类标准方案设计，包括立面、内装、景观、夜景、导向标识5大类效果类专项设计。结合不同的城市级别，万达广场的5大类效果类方案分为A-、B和B-三种标准。新开发的万达广场一旦确定立面，相应地，其内装、景观、导向标识、夜景也会确定，极大地缩短了前期的设计周期，为后续的快速发展创造了便利条件（图1~图4）。

5. 效果类标准样板库

为积极配合直投万达广场标准化工作，解决管控过程中施工总包、项目公司、设计供方的设计封样问题，提高工作效率及集采化程度，由万达商业规划院牵头建立"万达广场设计封样样板库"。样板库分实体库及电子库，全部材料统一编码，并与BIM编码系统一一对应，便于各方查询使用。

实体库以直投标准版万达广场各效果类专业封样材

standard plans. The module drawings are used for guiding departments of Wanda Group and designers to quickly select standard modules for design, thus enhancing standardization rate of direct invested Wanda Plazas and reaching Group's strategic goal for BIM Standardized Design.

What the atlas contains is the content that is frequently seen in direct invested Wanda Plazas' standard drawings and suitable for modularization design, such as stair, elevator, toilet and duct well, etc. To this end, it is easily applied by each department and very applicable to standard, near-standard and non-standard projects.

The standard module units are optional at scheme design phase, and used upon adjustment according to actual project situations at preliminary design and construction drawings design phases. The drawings encompass three categories, respectively being volume of architecture, volume of M&E, and volume of landscape.

Sky light roof of Wanda Plazas, for instance, has developed a technological process fitting factory production and prefabricated construction. It consists of seven systems, including envelop system, stress system, fire control system, sun-shading system, lighting system, hanging system and lightning-thunder proof system, each component's standardization design of which ultimately contributes to *Standard Design Atlas of Wanda Plazas Sky light Roof Industrialization*. Thanks to systematic research on the first three systems and the following industrialized production, fabricated installation and informatization management, the standardized design Modular construction of daylighting roof is made available. Till now, the daylighting roof has completed industrialization pilots and delivered remarkable outcomes.

4. SPECIAL STANDARD MODULARIZATION

The year 2015 witnessed the completion of Wanda Plazas' effect category scheme design, covering special design for five aspects, namely facade, interior design, landscape, nightscape and guidance signs. According to city levels, the five effect schemes are subject to three standards (A-, B and B-). This means once a facade of a Wanda Plaza is decided, its interior design, landscape, signage signs and nightscape will be determined accordingly, which greatly shortens the design cycle at the early phase and paves the way for subsequent rapid development (Fig.1-4).

5. EFFECT CATEGORY DESIGN PROTOTYPE LIBRARY

Wanda Commercial Planning Institute takes the initiative to establish a prototype library of Wanda Plazas' sealed design material samples, so as to coordinate with standardization of directly invested Wanda Plazas, address design-freeze stage material samples among main construction contractors, project companies and design providers, and to improve work efficiency and concentrated purchase degree. The prototype library consisting of a physical library and an electronic library have all materials encoded uniformly and corresponded to BIM encoding system for query.

(图1) 立面标准方案

(图2) 内装标准方案图

（图3）景观标准方案

（图4）夜景标准方案

（图5）实体样板库

料为主，采用通用材料分类体系，按专业分为建筑立面、内装、景观、夜景、导向标识等5大版块19类材料，共93种材料，覆盖100%直投项目设计封样。电子库中每一例设计封样材料，相关详细信息均可通过规划院网站、移动终端万达规划APP、二维码等查询（图5）。

三、"万达广场标准化设计"研发后续方向和展望

在现有工作成果的基础上，万达广场标准化设计研发工作将会进一步结合"BIM总发包管理"模式，研究BIM标准构件库、BIM标准模型、设计模块标准方案，预期设计模块标准方案累计标准化率达到70%。

设计模块标准方案的研发成功和应用，可以方便项目前期与政府沟通汇报，快速实现对潜在项目预期经济效益的预判，缩短项目前期测算和决策周期，使项目用地符合企业标准化发展要求；设计管控部门直接使用，大大缩短设计周期，提高效率；商管部门在拿到项目后，可先期与商户沟通，了解商户需求，更早实现商业业态落位；还可以通过BIM等先进手段，实现集采招标，工厂化加工，更高效快捷地完成项目建设。

开展"万达广场标准化设计"研发将万达十六年商业地产规划设计实践工作进行研究、概括、总结、提升，为万达商业地产适应不同城市、不同建筑标准的万达广场实现快速开发、快速设计、快速建造打下了坚实的技术基础。

Largely containing all-specialty selected material samples of directly invested standard Wanda Plazas, the physical library adopts general materials classification system and is divided into five segments and 19 material classes, totaling 93 kinds of materials and covering design-freeze material samples of 100% directly invested projects. Details of all selected design material samples in the E-library can be obtained through the Wanda Planning and Research Institute's website, mobile terminal Wanda Planning APP and QR code (Fig.5).

III. DIRECTION AND PROSPECTS OF STANDARDIZED DESIGN

On the basis of existing work results, R&D work on Standardized Design will carry out further research on the BIM standard component library, BIM standard model and design module standard scheme, combined with the model of BIM Overall Contracting Management. The design module standard scheme is expected to reach 70% in accumulative standardization rate.

Successful R&D and application of the design module standard schemes have the following significances. First, it facilitates communication and report with governmental authorities at the early phase and rapid expected economic benefit estimate for potential projects, shortens project analysis and decision cycle at the early phase, and enables land for projects to comply with enterprise standardization development demand. Second, it can be directly applied by the Design and Control Department to significantly quicken design cycle and improve efficiency. Third, after winning projects, the Commercial Management Department may communicate with merchants in advance to clear their demands and to make commercial activities established faster. Fourth, advanced technologies such as BIM can be used for concentrated procurement bidding and factory manufacturing, so as to complete project construction more efficiently.

R&D on Standardized Design has studied, summarized, generalized and upgraded Wanda's sixteen-year engagement in the commercial properties planning and design practice, laying solid foundation for rapid development, design and construction of Wanda Plazas of different cities and construction standards.

R&D AND APPLICATION OF "HUIYUN SYSTEM" V2.0
"慧云系统"（2.0版）的研发及应用

万达商业规划研究院副院长　方伟
万达商业规划研究院总工程师　范珑

在国际、国内信息化及智能化大潮下，万达集团自主研发了具有独立知识产权的"慧云智能化管理系统"（以下简称"慧云系统"）。"慧云系统"2013年试点成功，2014年全面推广。通过20多个项目的建设实践与商管运营检验，在设计建造和运营管理方面积累了大量的经验，2015年"慧云"研发团队秉承不断探索、追求创新的企业文化精神，进一步梳理运营管理需求，按照"实用"、"有用"的原则，经过"复盘、检查、调研、总结"，在确保不降低运营安全、保证运营品质的前提下，重新界定"慧云"功能划分，调整了"慧云"集成平台与"慧云"子系统的功能分布，调整了"慧云系统"具体操控方式，研发出"慧云系统"（2.0版）。新的"慧云系统"具有"现场实施更加简单，系统功能更加优化，考核标准更加严格"三大特点。

一、"慧云系统"（2.0版）现场实施更加简单

"慧云系统"包括5大管理职能、16个弱电子系统、监控设备3000多台、监控信息点位数万个，涉及弱电智能化、信息化、机电、消防等多个专业领域的设计和施工；而我国尚未制定这种跨学科、跨领域的设计与施工验收技术标准，跨系统深度集成在国际上也是新课题。"慧云系统"的实施对设计、施工和运营都提出了极高的要求，某种程度上超出了行业的整体能力。

基于上述原因，"慧云系统"（2.0版）保留核心功能，去除不重要功能，以此为原则调整了16个子系统与平台的集成方式。根据各子系统的操作特点与商管运营的使用需求，将4个系统以物理集成方式接入慧云机房独立操作，将6个系统通过页面嵌套接入"慧云系统"，保留了6个系统通过"数据交互"与"集成绑点"完全集成接入"慧云系统"。子系统集成方式调整后，最重要的"消防系统"独立运行，其操作显示屏设在离消防报警主机最近的操作台左侧，旁边是视频监控系统显示屏。这种布置便于管理员通过视频监控系统了解火灾报警现场情况、起身至消防报警主机操作等；在操作台上为日常运营管理使用最多的BA系统设置了独立的操作屏，其中包含了"暖通空调"和"给排水"两个系统；操作台最右侧是可以显示其余各子系统内容的综合显示屏，通

Under the trend of informatization and intellectualization at home and abroad, Wanda Group has independently researched and developed Huiyun Intelligent Management System (hereinafter referred to as "Huiyun System") with independent intellectual property rights. "Huiyun System" was successfully tested in 2013 and widely applied in 2014. Through the construction of more than 20 projects and operation & inspection of the Commercial Management Department, tremendous experience relating to design & construction and operation management has been accumulated. In 2015, adhering to the enterprise culture spirit of continuously exploring and pursuing innovation, "Huiyun System" R&D team has further organized demands for operation management. Following the practical and useful principle and after checking, inspection, research and summarization, the team has redefined functional division of "Huiyun System", adjusted the Huiyun integration platform, function distribution of Huiyun subsystem and specific control mode of "Huiyun System", without compromising on operation security and quality, and ultimately developed the "Huiyun System" V2.0. The new "Huiyun System" is characterized by easier onsite implementation, optimized system function and stricter assessment criterion.

I. EASIER ONSITE IMPLEMENTATION OF HUIYUAN SYSTEM V2.0

"Huiyun System" comprises of 5 management functions, 16 ELV subsystems, 3,000 plus monitoring equipment and tens of thousands of monitoring information points, and involves design and construction of multiple fields, such as ELV intelligentization and informatization, M&E and fire control. Yet such interdisciplinary and cross-domain design & construction acceptance technical standard has not been seen in China, and cross-system depth integration is also a new task in the world. Indeed, implementation of "Huiyun System" requests high demands on design, construction and operation, and to a certain degree, outperforms overall capacity of the industry.

Given the above-mentioned reasons, "Huiyun System" V2.0 pursues the principle of keeping core functions and removing unimportant ones to adjust integration mode of its 16 subsystems and platforms. According to the operation features of the subsystems and operation use requirements of the Commercial Management Department, four subsystems are connected to Huiyun plant room in a physical integration mode for independent operation, six subsystems are connected to "Huiyun System" through embedded pages, and the remaining six subsystems are fully integrated through data interaction and integrated binding points and then connected to "Huiyun System". Such adjustment is followed by independent operation of the crucial fire control system, the operation display screen of which is on the left of the operation desk nearest to fire alarm host, and beside video monitoring system display screen. With such placement, managers can be informed of fire alarm site condition via the monitoring system and operate fire alarm host. The operation desk is provided with an independent operation screen for

（图1）"慧云系统"（2.0版）主界面

过KVM设备实现了多个系统共用一台显示器的功能，通过对"慧云"平台与子系统的功能重新分配，实现了在一个操作台上监控所有机电系统。优化调整后的"慧云系统"在"慧云"平台上集成"绑点"的监控信息点位数量减少了60%，大大简化了现场实施难度。升级后的"慧云系统"（2.0版）主界面如图1所示，"慧云系统"（2.0版）架构如图2所示，"慧云系统"（2.0版）操作台如图3所示。

二、"慧云系统"（2.0版）系统功能更加优化

根据运行管理人员的使用需求，"慧云"（2.0版）在原设计标准的基础上新增功能75项，删除功能25项，修改功能52项。

针对各类型不同级别的报警信息，为了让使用人员能够快速获取系统运行报警状态、定位故障信息和便捷的报警处理流程，"慧云系统"（2.0版）将设备运行故障与设备运行参数报警信息区分开，报警屏采用了分栏显示的方式。同时报警管理功能细分为实时报警、历史报警和报警定义三部分（图4）。

在实时报警中可以直接浏览到报警产生原因、报警实时数据等关键参数，以及报警处理预案和进行应答处理状态；为避免信息干扰，同一报警信息只显示一条，不再重复显示。当实时报警窗口中的报警条目被处理或报警信息复位后，该条报警则自动归档到历史报警窗口中。在"慧云"平台可定义报警信息，报警定义的信息包括"报警描述"、"报警类型"和"报警

BA system frequently serving daily operation management, including HVAC and Plumbing subsystems. On the most right side of operation desk stands the integrated display screen, which is applied to multiple subsystems via KVM equipment. After functional redistribution for Huiyun platform and subsystems, M&E subsystem can be monitored through one operation desk. The optimized and adjusted "Huiyun System" has its monitoring information points of integrated binding points on Huiyun platform slashed 60%, which dramatically reduces onsite implementation difficulty. Main interface and architecture of the upgraded "Huiyun System" V2.0 are shown in Fig.1, Fig.2, Fig.3 respectively.

II. OPTIMIZED SYSTEM FUNCTION OF "HUIYUN SYSTEM" V2.0

In response to use requirements of operation managers, "Huiyun System" V2.0 has added 75 functions, omitted 25 functions and modified 52 functions based on the original design standard.

Directed at alarm information of different types and levels, "Huiyun System" V2.0 has separated equipment operation fault from equipment operation parameter alarm information, and alarm screen is displayed in columns. In this manner, users can promptly attain system operation alarm status, position fault message and access convenient alarm processing procedure. And alarm management functions are classified into real-time alarm, historical alarm and alarm definition (Fig.4).

In the real-time alarm, key parameters such as cuause of alarm and real-time alarm data, alarm processing plan and reply processing status can be directly viewed. To avoid information interference, the same alarm information is shown once only and not repeated. After alarm entry on real-time alarm window has been processed or alarm information has been reset, this piece of alarm is automatically archived into historical alarm window. Huiyun platform can define such alarm information as Alarm Description, Alarm Type and Alarm Level, etc. "Huiyun System" V2.0 mainly includes

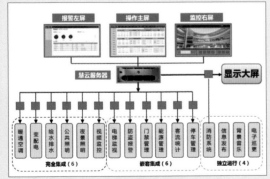

（图2）"慧云系统"（2.0版）架构示意图

（图3）"慧云系统"操作台

(图4)"慧云系统"(2.0版)报警屏分栏显示

等级"等。"慧云系统"(2.0版)的报警类型主要包括"设备报警"和"运行报警"两类。"设备报警"为设备故障相关报警,通常需运行人员现场核查;如确认为故障后,则启动相关的维保工作。"运行报警"为实际运行过程中发生的报警,也需运行人员现场核查,确认报警后需调整相关运行方式。

根据运行管理人员实际工作需求,"慧云系统"(2.0版)升级报表管理功能,以电子表格的形式输出各运行设备的运行参数,包括中压配电,锅炉,冷冻、冷却水泵、冷却塔、冷水机组、冷站系统、送排风机、新风机组、组合式空调机组等的运行参数数据。对于每一种设备的报表,"慧云系统"集成平台报表工具均可定时生成设备运行记录报表文件,供用户随时查看和下载。减轻了管理人员每天大量手动抄表的工作量,有效地提高了工作效率及工作质量。

三、"慧云系统"(2.0版)考核标准更加严格

与设计标准同时修订的还有"'慧云系统'验收标准"。"慧云系统"(2.0版)"验收标准"对工程完成度考核、"慧云系统"试运行考核及"慧云系统"联合验收都提出了更严格的要求。"工程完成度考核"在项目开业后30天完成,由商管总部和规划设计系统分别派人参加,现场检查67项内容,涉及数百个信息点位,及格分数线从原来的70分提高到85分;对"慧云系统"试运行节点的要求是对各系统进行全覆盖检测,涉及3万多个信息点位,所有功能与信息点位错误率不得高于2%为合格;联合验收由商管总部、信息

two types of alarm, being Equipment Alarm and Operation Alarm. The former refers to equipment fault related alarms, which usually require operators' onsite check and will initiate related maintenance work after the fault is confirmed. The latter refers to alarms in the process of operation, which also requires onsite check and adjustment of related operation mode after the alarm is confirmed.

Considering the actual work demands of operation personnel, "Huiyun System" V2.0 has upgraded report management function to output operating parameters of each operation equipment, including the medium voltage power distribution, boiler, refrigeration, cooling water pump, cooling tower, water chilling unit, chiller plant system, supply and exhaust fan, fresh air units and combined air conditioning units, among others. For any equipment report, reporting tools of Huiyun System integration platform can generate operation record report documents regularly for users to view and download at any time. The upgraded function relieves managers' from tremendous daily meter-reading work, and effectively improves the work efficiency and work quality.

III. STRICTER ASSESSMENT CRITERION OF "HUIYUN SYSTEM" V2.0

Acceptance criteria of "Huiyun System" V2.0 have been revised along with design criteria, and posed more rigorous requirements on project completion assessment and commissioning assessment and joint acceptance of "Huiyun System". Project completion assessment is required to be finished 30 days after the project is opened for business, attended by personnel from the Commercial Management Headquarters and Planning & Design Unit, and covers 67 onsite inspection items and hundreds of information points, with passing score raised from 70 to 85. Commissioning assessment requires detection for all systems and involves 30,000 plus information points. If error rate of all functions and information points is no more than 2%, it is deemed to be qualified. Joint acceptance is separately scored by the Commercial Management Headquarters, Information Center and Safety Supervision Department, and involves 121 onsite

中心和安监部负责，各部门独立打分，现场检查121项内容，涉及近千个信息点位，及格分数线也从85分提高到95分。另外，"工程完成度考核"与"慧云系统"联合验收标准还增加了若干条"一票否决"项，对严重影响后期运行的问题实行一票否决。

"慧云系统"（2.0版）标准于2015年7月中旬颁布，至2015年底，已经有29个改造项目通过"慧云系统"（2.0版）验收标准、所有新建项目通过不同阶段的考核验收。数十个项目顺利通过最严格考核验收标准的事实说明，采用新架构实施的"慧云系统"均能达到新验收标准的要求，系统存在的问题大幅度降低。"慧云系统"（2.0版）在不降低运行安全性、不改变操作便利性的前提下优化了系统，大大降低了现场实施难度，提高了验收的考核标准，能够更好地满足商管的使用要求。

四、"慧云系统"后续工作展望

在"慧云系统"（2.0版）的基础上，2015年完成了"慧云系统"（3.0版）的研发立项，2016年将研发"慧云系统"（3.0版）。"慧云系统"（3.0版）是一种"云架构"方案，具有云端部署、总部集中运维、统一数据备份、版本统一发布等多项优势，使"慧云系统"的管控及升级更加方便。同时要求各地项目数据实时上传总部，可对业务数据进行大数据分析，为深度挖掘数据价值创造了有利条件。"慧云系统"（3.0版）建成后将成为全球最大规模的"云架构"智能化管理系统。

"慧云系统"的不断完善，使万达广场在运营管理方面比同行业竞争者拥有更多的技术优势，有助于降低人工成本、保证运行品质、降低运行能耗，有利于万达广场资产的保值、升值。充分挖掘"慧云系统"的数据价值，可以进一步优化设计方案，降低建造成本，提升运营管理能力，设计、建造、运营形成良性循环，有助于万达广场品牌的推广，为集团战略转型的顺利实施保驾护航。

inspection items and nearly thousand points, with passing score raised from 85 to 95. In addition, project completion assessment and joint acceptance of "Huiyun System" have introduced a number of one-vote veto options against those issues seriously affecting operation at later phase.

In mid-July of 2015, "Huiyun System" V2.0 standard was issued. And by the end of the year, 29 renovation projects have passed its acceptance criteria and all newly built projects have passed assessment acceptance at corresponding phases. The fact that dozens of projects have passed the strictest assessment acceptance criteria proves that, the new structure-based "Huiyun System" can satisfy new acceptance criteria requirements and existing system problems are largely reduced. Without compromising on operation safety and operation convenience, "Huiyun System" V2.0 has optimized system, significantly reduced onsite implementation difficulty and improved assessment criteria for acceptance, better catering for operation requirements of the Commercial Management Department.

IV. PROSPECTS FOR "HUIYUN SYSTEM"

In 2015, Wanda has initiated a project for "Huiyun System" V3.0 based on the Version 2.0. In 2016, Wanda will develop Version 3.0, a Cloud-based scheme. It has the advantages of cloud deployment, centralized operation and maintenance by the headquarters, unified data backup and unified release of edition, and facilitates the control and upgrading of "Huiyun System". In addition, the new scheme stipulates that project companies in various regions shall upload real-time project data to the headquarters; therefore, big data analysis can be carried out based on business data, which shall create favorable conditions for the excavation of the data value in depth. Upon its completion, "Huiyun System" V3.0 is set to be the world's largest cloud-based intelligent management system.

With the consistent improvement of "Huiyu System", Wanda Plaza shall have more technical advantages over its competitors in the same industry in the aspect of operational management, which will not only help cut down manpower costs, ensure operating quality and reduce operating energy consumption, but also make for Wanda Plaza assets' maintenance and increase of value. With full exploitation on the value of the data collected by "Huiyun System", favorable results of design plan further optimization, construction costs reduction, promotion of operational management ability and formation of a virtuous circle among design, construction and operation can be achieved, which shall contribute to the brand promotion of Wanda Plaza and safeguard the smooth implementation of the Group's Asset-light Strategy.

F

PROJECT INDEX
项目索引

WANDA COMMERCIAL PLANNING 2015

INDEX OF WANDA PLAZAS
万达广场索引

P01 TAIYUAN LONGFOR WANDA PLAZA 太原龙湖万达广场

大商业施工图设计单位	太原市建筑设计研究院
外立面设计单位	北京华雍汉维建筑咨询有限公司
内装设计单位	北京清尚环艺建筑设计院有限公司
景观设计单位	中国建筑设计院有限公司
导向标识设计单位	北京艺同博雅企业形象设计有限公司
夜景照明设计单位	栋梁国际照明设计（北京）中心有限公司
弱电智能化设计单位	上海智信世创智能系统集成有限公司
外幕墙深化设计单位	北京市金星卓宏幕墙工程有限公司

P02 CHONGQING BA'NAN WANDA PLAZA 重庆巴南万达广场

大商业施工图设计单位	中国建筑西南设计研究院有限公司
外立面设计单位	卡斯帕建筑设计咨询（上海）有限公司
内装设计单位	北京清尚环艺建筑设计院有限公司
景观设计单位	中国建筑设计研究院
导向标识设计单位	北京视域四维城市导向系统规划设计有限公司
夜景照明设计单位	深圳千百辉照明工程有限公司
弱电智能化设计单位	北京国安电气有限责任公司
外幕墙深化设计单位	北京市金星卓宏幕墙工程有限公司

P03 DALIAN KAIFAQU WANDA PLAZA 大连开发区万达广场

大商业施工图设计单位	大连都市发展建筑设计有限公司
外立面设计单位	亚瑞建筑设计有限公司
内装设计单位	北京伊品设计有限公司
景观设计单位	北京中建城市设计院有限公司上海分公司
导向标识设计单位	北京视域四维城市导向系统规划设计有限公司
夜景照明设计单位	千百辉照明工程有限公司
弱电智能化设计单位	大连理工科技有限公司
外幕墙深化设计单位	北京国科天创建筑设计院有限责任公司

P04 SHANGHAI JINSHAN WANDA PLAZA 上海金山万达广场

大商业施工图设计单位	中国建筑上海设计研究院有限公司
外立面设计单位	上海鼎实建筑设计有限公司
内装设计单位	北京市建筑装饰设计院有限公司
景观设计单位	深圳文科园林股份有限公司
导向标识设计单位	北京艺同博雅企业形象设计有限公司
夜景照明设计单位	深圳普莱思照明设计顾问有限责任公司
弱电智能化设计单位	上海智信世创智能系统集成有限公司
外幕墙深化设计单位	厦门开联装饰工程有限公司

P05 GUANGZHOU LUOGANG WANDA PLAZA 广州萝岗万达广场

大商业施工图设计单位	广州宝贤华翰建筑工程设计有限公司
外立面设计单位	北京华雍汉维建筑咨询有限公司
内装设计单位	广东省集美设计工程有限公司
景观设计单位	深圳文科园林股份有限公司
导向标识设计单位	北京视域四维城市导向系统规划设计有限公司
夜景照明设计单位	深圳市千百辉照明工程有限公司
弱电智能化设计单位	上海中电电子系统工程有限公司
外幕墙深化设计单位	厦门开联装饰工程有限公司

P06 DONGGUAN HOUJIE WANDA PLAZA 东莞厚街万达广场

大商业施工图设计单位	深圳市华森建筑工程咨询有限公司
外立面设计单位	北京中外建筑设计有限公司
内装设计单位	广东省集美设计工程有限公司
景观设计单位	上海帕莱登建筑景观咨询有限公司
导向标识设计单位	北京视域四维城市导向系统规划设计有限公司
夜景照明设计单位	深圳市粤大明智慧照明科技有限公司
弱电智能化设计单位	上海智信世创智能系统集成有限公司
外幕墙深化设计单位	北京和平幕墙工程有限公司

P07 LIUZHOU CHENGZHONG WANDA PLAZA 柳州城中万达广场

大商业施工图设计单位	广州宝贤华瀚建筑工程设计有限公司
外立面设计单位	上海对外建设建筑设计有限公司
内装设计单位	北京清尚建筑设计研究院有限公司
景观设计单位	北京中建建筑设计院有限公司上海分公司
导向标识设计单位	北京艺同博雅企业形象设计有限公司
夜景照明设计单位	深圳市千百辉照明工程有限公司
弱电智能化设计单位	北京国安电气有限责任公司
外幕墙深化设计单位	厦门开联装饰工程有限公司

P08 GUILIN GAOXIN WANDA PLAZA 桂林高新万达广场

大商业施工图设计单位	大连市建筑设计研究院有限公司
外立面设计单位	青岛北洋建筑设计有限公司
内装设计单位	中艺建筑装饰有限公司
景观设计单位	中国建筑设计研究院
导向标识设计单位	北京艺同博雅企业形象设计有限公司
夜景照明设计单位	北京鱼禾光环境设计有限公司
弱电智能化设计单位	北京国安电气有限责任公司
外幕墙深化设计单位	深圳金粤幕墙装饰工程有限公司杭州分公司

P09 GUANGZHOU NANSHA WANDA PLAZA 广州南沙万达广场

大商业施工图设计单位	广州市设计院
外立面设计单位	豪斯泰勒张思图闻德建筑设计咨询（上海）有限公司
内装设计单位	广州市城市组设计有限公司
景观设计单位	福建泛亚远景环境设计工程有限公司
导向标识设计单位	北京艺同博雅企业形象设计有限公司
夜景照明设计单位	栋梁国际照明设计（北京）中心有限公司
外幕墙深化设计单位	北京市金星卓宏幕墙工程有限公司
弱电智能化设计单位	南京熊猫信息产业有限公司
外幕墙深化设计单位	厦门嘉福幕墙铝窗有限公司

P10 NANNING ANJI WANDA PLAZA 南宁安吉万达广场

大商业施工图设计单位	悉地国际设计顾问（深圳）有限公司
外立面设计单位	青岛北洋建筑设计有限公司
内装设计单位	广东省集美设计工程有限公司
景观设计单位	深圳市铁汉生态环境股份有限公司
导向标识设计单位	北京视域四维城市导向系统规划设计有限公司
夜景照明设计单位	北京三色石环境艺术有限公司
弱电智能化设计单位	北京时代凌宇科技股份有限公司
外幕墙深化设计单位	厦门开联装饰工程有限公司

P11 SICHUAN GUANGYUAN WANDA PLAZA 四川广元万达广场

大商业施工图设计单位	中国中轻国际工程有限公司
外立面设计单位	华凯派特建筑设计（北京）有限公司
内装设计单位	北京市建筑装饰设计院有限公司
景观设计单位	华汇工程设计集团股份有限公司
导向标识设计单位	北京广育德视觉技术有限公司
夜景照明设计单位	深圳金达照明股份有限公司
弱电智能化设计单位	北京国安电气有限责任公司
外幕墙深化设计单位	深圳蓝绿建建设集团股份有限公司

P12 NANTONG GANGZHA WANDA PLAZA 南通港闸万达广场

大商业施工图设计单位	北京东方国兴建筑设计有限公司
外立面设计单位	北京东方国兴建筑设计有限公司
内装设计单位	中国中建设计集团有限公司
景观设计单位	上海兴田建筑工程设计事务所
导向标识设计单位	北京艺同博雅企业形象设计有限公司
夜景照明设计单位	上海易照景观设计有限公司
弱电智能化设计单位	上海中电电子系统工程有限公司
外幕墙深化设计单位	上海旭密林幕墙有限公司

P13
XISHUANGBANNA WANDA PLAZA
西双版纳万达广场

大商业施工图设计单位	广东省建筑设计研究院
外立面设计单位	北京五合国际建筑设计咨询有限公司
内装设计单位	北京市建筑装饰设计院有限公司
景观设计单位	上海帕莱登建筑景观咨询有限公司
导向标识设计单位	北京艺同博雅企业形象设计有限公司
夜景照明设计单位	北京三色石环境艺术有限公司
弱电智能化设计单位	上海智信世创智能系统集成有限公司
外幕墙深化设计单位	深圳蓝波绿建集团股份有限公司

P14
TAI'AN WANDA PLAZA
泰安万达广场

大商业施工图设计单位	悉地国际建筑顾问（深圳）有限公司
外立面设计单位	欧创建筑联合事务所有限公司
内装设计单位	深圳三九装饰工程有限公司
景观设计单位	华汇工程设计集团股份有限公司
导向标识设计单位	北京艺同博雅企业形象设计有限公司
夜景照明设计单位	深圳市标美照明设计工程有限公司
弱电智能化设计单位	南京熊猫信息产业有限公司
外幕墙深化设计单位	深圳市新山幕墙技术咨询有限公司

P15
DEZHOU WANDA PLAZA
德州万达广场

大商业施工图设计单位	北京市东方国兴建筑设计院有限公司
外立面设计单位	上海新外建工程设计与顾问有限公司
内装设计单位	上海帕莱登建筑景观咨询有限公司
景观设计单位	上海华东建筑设计研究院
导向标识设计单位	北京清尚环艺建筑设计有限公司
夜景照明设计单位	北京鱼禾光环境设计有限公司
弱电智能化设计单位	北京国安电气有限责任公司
外幕墙深化设计单位	北京市金星卓宏幕墙工程有限公司

P16
DONGYING WANDA PLAZA
东营万达广场

大商业施工图设计单位	悉地国际设计顾问（深圳）有限公司
外立面设计单位	思邦建筑设计咨询（上海）有限公司
内装设计单位	北京市建筑装饰设计院有限公司
景观设计单位	上海兴田建筑工程设计事务所
导向标识设计单位	北京艺同博雅企业形象设计有限公司
夜景照明设计单位	北京三色石环境艺术有限公司
弱电智能化设计单位	上海中电电子系统工程有限公司
外幕墙深化设计单位	北京市金星卓宏幕墙工程有限公司

P17
HUANGSHI WANDA PLAZA
黄石万达广场

大商业施工图设计单位	江西省建筑设计研究院
外立面设计单位	华凯帕特建筑设计（北京）有限公司
内装设计单位	北京清尚建筑设计研究院有限公司
景观设计单位	北京中建建筑设计院有限公司上海分公司
导向标识设计单位	北京视域四维城市导向系统规划设计有限公司
夜景照明设计单位	福建super大建筑设计有限公司
弱电智能化设计单位	北京益泰牡丹电子工程有限责任公司
外幕墙深化设计单位	深圳金粤幕墙装饰工程有限公司

P18
ZHEJIANG JIAXING WANDA PLAZA
浙江嘉兴万达广场

大商业施工图设计单位	华汇工程设计集团股份有限公司
外立面设计单位	青岛北洋建筑设计有限公司
内装设计单位	上海浦东建筑设计研究院有限公司
景观设计单位	华东建筑设计研究院有限公司
导向标识设计单位	北京广育德视觉技术股份有限公司
夜景照明设计单位	上海易阳景观设计有限公司
弱电智能化设计单位	上海智信世创智能系统集成有限公司
外幕墙深化设计单位	北京市金星卓宏幕墙工程有限公司

P19
SUZHOU WUZHONG WANDA PLAZA
苏州吴中万达广场

大商业施工图设计单位	苏州设计研究院股份有限公司
外立面设计单位	上海对外建筑建设集团
内装设计单位	中国建筑设计研究院
景观设计单位	深圳文科园林股份有限公司
导向标识设计单位	北京艺同博雅企业形象设计有限公司
夜景照明设计单位	深圳市千百辉照明工程有限公司
弱电智能化设计单位	上海中电电子系统工程有限公司
外幕墙深化设计单位	北京市金星卓宏幕墙工程有限公司

P20
FUYANG YINGZHOU WANDA PLAZA
阜阳颍州万达广场

大商业施工图设计单位	中国建筑上海设计研究院有限公司
外立面设计单位	上海思纳建筑设计集团
内装设计单位	上海帕莱登建筑景观咨询有限公司
景观设计单位	华东建筑设计研究院有限公司
导向标识设计单位	北京清华同衡规划设计研究院有限公司
夜景照明设计单位	上海易景观设计有限公司
弱电智能化设计单位	北京国安电气有限责任公司
外幕墙深化设计单位	深圳市新山幕墙技术咨询有限公司

P21
NEIJIANG WANDA PLAZA
内江万达广场

大商业施工图设计单位	重庆市设计院
外立面设计单位	艾奕康建筑设计（深圳）有限公司上海分公司
	北京赫斯科建筑设计咨询有限公司
内装设计单位	北京市建筑装饰设计院有限公司
景观设计单位	华汇工程设计集团股份有限公司
导向标识设计单位	北京艺同博雅企业形象设计有限公司
夜景照明设计单位	北京清华同衡规划设计研究院有限公司
弱电智能化设计单位	大连理工科技有限公司
外幕墙深化设计单位	深圳金粤幕墙装饰工程有限公司

P22
QIQIHAR WANDA PLAZA
齐齐哈尔万达广场

大商业施工图设计单位	哈尔滨工业大学建筑设计研究院有限公司
外立面设计单位	青岛腾远设计事务所有限公司
内装设计单位	北京市建筑装饰设计院有限公司
景观设计单位	深圳文科园林股份有限公司
导向标识设计单位	北京视域四维城市导向系统规划设计有限公司
夜景照明设计单位	深圳市标美照明设计工程有限公司
弱电智能化设计单位	大连理工科技有限公司
外幕墙深化设计单位	深圳市新山幕墙技术咨询有限公司

P23
ANYANG WANDA PLAZA
安阳万达广场

大商业施工图设计单位	机械工业第六设计研究院有限公司
外立面设计单位	华凯国际（上海）建筑设计咨询有限公司
内装设计单位	上海浦东建筑设计研究院有限公司
景观设计单位	华东建筑设计研究院有限公司
导向标识设计单位	北京艺同博雅企业形象设计有限公司
夜景照明设计单位	深圳市千百辉照明工程有限公司
弱电智能化设计单位	华体集团有限公司
外幕墙深化设计单位	深圳市三鑫幕墙工程有限公司

P24
WEINAN WANDA PLAZA
渭南万达广场

大商业施工图设计单位	北京新纪元建筑工程设计有限公司
外立面设计单位	上海思纳建筑规划设计有限公司
内装设计单位	北京市建筑装饰设计院有限公司
景观设计单位	福建泛亚远景环境设计工程有限公司
导向标识设计单位	北京艺同博雅企业形象设计有限公司
夜景照明设计单位	上海译格照明设计有限公司
弱电智能化设计单位	上海中电电子系统工程有限公司
外幕墙深化设计单位	深圳蓝波绿建集团股份有限公司

P25
YINGKOU WANDA PLAZA
营口万达广场

大商业施工图设计单位	大连市建筑设计研究院有限公司
外立面设计单位	青岛北洋建筑设计有限公司
内装设计单位	中艺建筑装饰有限公司
景观设计单位	中国建筑设计研究院
导向标识设计单位	北京艺同博雅企业形象设计有限公司
夜景照明设计单位	北京鱼禾光环境设计有限公司
弱电智能化设计单位	北京国安电气有限公司
外幕墙深化设计单位	深圳金粤幕墙装饰工程有限公司杭州分公司

P26
JIAMUSI WANDA PLAZA
佳木斯万达广场

大商业施工图设计单位	哈尔滨工业大学建筑设计研究院
外立面设计单位	北京市金星卓宏幕墙工程有限公司
内装设计单位	北京市建筑装饰设计院有限公司
景观设计单位	北京中建建筑设计院有限公司
导向标识设计单位	北京艺同博雅企业形象设计有限公司
夜景照明设计单位	北京鱼禾光环境设计有限公司
弱电智能化设计单位	北京国安电气有限责任公司
外幕墙深化设计单位	秦皇岛渤海铝幕墙装饰工程有限公司

INDEX OF WANDA HOTELS
万达酒店索引

H01
WANDA REIGN CHENGDU
成都万达瑞华酒店

土建施工图设计单位	中国建筑西南设计研究院有限公司
外立面设计单位	德国KSP建筑设计事务所
内装设计单位	万达酒店设计研究院
景观设计单位	上海陆道工程设计管理股份有限公司
导向标识设计单位	北京艺同博雅企业形象设计有限公司
夜景照明设计单位	上海译格照明设计有限公司
机电智能化设计单位	万达酒店设计研究院
外幕墙深化设计单位	北京和平幕墙工程有限公司

H02
WANDA VISTA RESORT XISHUANGBANNA
西双版纳万达文华度假酒店

土建施工图设计单位	大连市建筑设计研究院有限公司
外立面设计单位	北京欧安地合众建筑设计顾问有限公司
内装设计单位	万达酒店设计研究院
景观设计单位	上海帕莱登建筑景观咨询有限公司（公区及套房区）
	上海陆道工程设计管理股份有限公司（普通客房区）
导向标识设计单位	北京广育德视觉技术股份有限公司
夜景照明设计单位	上海译格照明设计有限公司
机电智能化设计单位	万达酒店设计研究院
外幕墙深化设计单位	深圳蓝波绿建集团股份有限公司

H03
WANDA VISTA HOHHOT
呼和浩特万达文华酒店

土建施工图设计单位	内蒙古工大建筑设计有限公司
外立面设计单位	北京奥思德建筑设计有限公司
内装设计单位	万达酒店设计研究院
景观设计单位	宝佳丰(北京)国际建筑景观规划设计有限公司
导向标识设计单位	陕西美心环境标识有限公司
夜景照明设计单位	上海译格照明设计有限公司
机电智能化设计单位	万达酒店设计研究院
外幕墙深化设计单位	北京市金星卓宏幕墙工程有限公司

H04
WANDA DOUBLETREE RESORT BY HILTON HOTEL XISHUANGBANNA
西双版纳万达希尔顿逸林度假酒店

土建施工图设计单位	奥意建筑工程设计有限公司
外立面设计单位	北京欧安地合众建筑设计顾问有限公司
内装设计单位	万达酒店设计研究院
景观设计单位	深圳致意思维筑景设计有限公司
导向标识设计单位	北京广育德视觉技术股份有限公司
夜景照明设计单位	上海译格照明设计有限公司
机电智能化设计单位	万达酒店设计研究院
外幕墙深化设计单位	深圳蓝波绿建集团股份有限公司

H05
WANDA REALM LIUZHOU
柳州万达嘉华酒店

土建施工图设计单位	广州宝贤华瀚建筑工程有限公司
外立面设计单位	上海对外建设设计有限公司
内装设计单位	万达酒店设计研究院
景观设计单位	北京中建建筑设计院上海分公司
导向标识设计单位	北京广育德视觉技术有限公司
夜景照明设计单位	深圳市千百辉照明工程有限公司
机电智能化设计单位	万达酒店设计研究院
外幕墙深化设计单位	厦门开联装饰工程有限公司

H06
WANDA REALM TAI'AN
泰安万达嘉华酒店

土建施工图设计单位	悉地国际设计顾问(深圳)有限公司
外立面设计单位	夏邦杰建筑设计咨询(上海)有限公司
内装设计单位	万达酒店设计研究院
景观设计单位	华汇工程设计集团股份有限公司
导向标识设计单位	北京艺同博雅企业形象设计有限公司
夜景照明设计单位	深圳市标美照明设计工程有限公司
机电智能化设计单位	万达酒店设计研究院
外幕墙深化设计单位	深圳市新山幕墙技术咨询有限公司

H07
WANDA REALM HUANGSHI
黄石万达嘉华酒店

土建施工图设计单位	江西省建筑设计研究院
外立面设计单位	华凯帕特建筑设计（北京）有限公司
内装设计单位	CCD 香港郑中设计事务所
景观设计单位	北京中建建筑设计有限公司上海分公司
导向标识设计单位	北京视域四维城市导向系统规划设计有限公司
夜景照明设计单位	福建福大建筑设计有限公司
机电智能化设计单位	中信建筑设计研究总院有限公司
外幕墙深化设计单位	深圳金粤幕墙装饰工程有限公司

H08
WANDA REALM ANYANG
安阳万达嘉华酒店

土建施工图设计单位	机械工业第六设计研究院有限公司
外立面设计单位	华凯国际（上海）建筑设计咨询有限公司
内装设计单位	万达酒店设计研究院
	Leo International Design Cooup Co., Ltd.
景观设计单位	华东建筑设计研究院有限公司
导向标识设计单位	北京艺同博雅企业形象设计有限公司
夜景照明设计单位	深圳市千百辉照明工程有限公司
机电智能化设计单位	深圳奥意建筑工程设计有限公司
外幕墙深化设计单位	西安高科幕墙门窗有限公司幕墙设计院

H09
WANDA REALM GUANGYUAN
广元万达嘉华酒店

土建施工图设计单位	中国中轻国际工程有限公司
外立面设计单位	华凯派特建筑设计（北京）有限公司
内装设计单位	万达酒店设计研究院
景观设计单位	深圳市黎奥室内设计有限公司
导向标识设计单位	华汇工程设计集团股份有限公司
夜景照明设计单位	北京广育德视觉技术股份有限公司
机电智能化设计单位	深圳金达照明股份有限公司
外幕墙深化设计单位	深圳奥意建筑工程设计有限公司
	深圳蓝波绿建集团股份有限公司

H10
WANDA REALM NEIJIANG
内江万达嘉华酒店

土建施工图设计单位	重庆市设计院
外立面设计单位	北京赫斯科建筑设计咨询有限公司
内装设计单位	苏州金螳螂建筑装饰股份有限公司
景观设计单位	华汇工程设计集团股份有限公司
导向标识设计单位	北京艺同博雅企业形象设计有限公司
夜景照明设计单位	北京清华同衡规划设计研究院有限公司
机电智能化设计单位	深圳奥意建筑工程设计有限公司
外幕墙深化设计单位	深圳金粤幕墙装饰工程有限公司

H11
WANDA REALM DONGYING
东营万达嘉华酒店

土建施工图设计单位	悉地国际设计顾问（深圳）有限公司
外立面设计单位	思邦建筑设计咨询（上海）有限公司
内装设计单位	万达酒店设计研究院
景观设计单位	上海兴田建筑工程设计事务所
导向标识设计单位	北京艺同博雅企业形象设计有限公司
夜景照明设计单位	北京三色石环境艺术设计有限公司
机电智能化设计单位	万达酒店设计研究院
外幕墙深化设计单位	北京市金星卓宏幕墙工程有限公司

H12
WANDA REALM FUYANG
阜阳万达嘉华酒店

土建施工图设计单位	中国建筑上海设计研究院有限公司
外立面设计单位	上海思纳建筑设计有限公司
内装设计单位	万达酒店设计研究院
景观设计单位	华东建筑设计研究院有限公司
导向标识设计单位	北京清华同衡规划设计研究院有限公司
夜景照明设计单位	上海易照景观设计有限公司
机电智能化设计单位	万达酒店设计研究院
外幕墙深化设计单位	深圳市新山幕墙技术咨询有限公司

H13
WANDA CROWNE PLAZA RESORT XISHUANGBANNA PARKVIEW
西双版纳万达皇冠假日度假酒店

土建施工图设计单位	奥意建筑工程设计有限公司
外立面设计单位	北京欧安地合众建筑设计顾问有限公司
内装设计单位	万达酒店设计研究院
景观设计单位	深圳致道思维景筑设计有限公司
导向标识设计单位	北京广育德视觉技术股份有限公司
夜景照明设计单位	上海译格照明设计有限公司
机电智能化设计单位	万达酒店设计研究院
外幕墙深化设计单位	深圳蓝波绿建集团股份有限公司

2015
万达商业规划
持 有 类 物 业

WANDA COMMERCIAL PLANNING 2015
PROPERTIES FOR HOLDING

朱其玮　叶宇峰　冯腾飞　刘冰　兰峻文　康斌　刘婷　马红　张琳
范珑　孙培宇　刘江　袁志浩　张宝鹏　王雪松　李海龙　耿大治　黄勇
蓝毅　沈余　李文娟　侯卫华　杨旭　屈娜　孙佳宁　徐立军　周澄　张涛
闫红伟　李江涛　石路也　张立峰　钟光辉　张洋　孟祥宾　章宇峰
吴绿野　徐小莉　张振宇　谷建芳　李彬　毛晓虎　万志斌　吴迪　李斌
党恩　高振江　孙海龙　黄引达　孙辉　齐宗新　杨艳坤　程欢　邓金坷
张宁　王群华　李昕　罗沁　刘佩　曹春　兰勇　李舒仪　曹亚星　陈维
吴晓璐　刘易昆　冯志红　都晖　陈杰　李小强　葛宁　张鹏翔　李洪涛
虞朋　吕鲲　康宇　王治天　董根泉　任腾飞　王吉　宫赫谣　沈文忠
张震　刘洋　方文奇　郭晨光　胡存珊　李志华　宋锦华　朱迪　秦鹏华
刘锋　凌峰　张堃　李涛　李明泽　陈勇　赵洪斌　赵海滨　李浩　刘志业
王玉龙　冯童　黄路　张剑锋　周德　李易　段堃　闫颇　朱欢　唐杰
刘潇　熊厚　王静　董明海　黄川东　黄涛　王昉　李捷　关发扬　庞博
任意刚　赵青扬　张争　张志斌　罗贤君　郭杨　李梦雷　赵陨　杨春龙
路滨　汪家绍　王少雷　张顺　白宝伟　贺明　庞庆　卫立新　冯晓芳
何志勇　宋永成　郭宇飞　卜少乐　刘海洋　韩冰　高峰　王睿麟　王凯
张佳　曹国峰　李常春　王永磊　于崇　张覻　杨汉国　赵剑利　王文广
罗冲　王权　方伟　刘俊　陈海亮　晁志鹏　邹洪　张德志　陈志强
周永会　桑伟　陈涛　高霞　王俊君　王清文　吴凡　张黎明　谭瑶
路清淇　刘安　陈晓州　汤英杰　周明　崔勇　陈理力　刘昕　韦云
杨娜　杨华　王朝志　罗琼　刘晓波　张烁君　魏大强　柏久绪　闵盛勇
蒲峰　朱广宇　汤钧　主佳　李扬　孟昆廷　李丽　梁超　钟文渊
王云　王奕　张雁翔　余斌　陈玼潭　王进纯　马雪健　张述杰
宋雷　李万勇　耿磊　王翔　马长宁　姚建刚　李韦达　马刚　杨琳
王连发　刘向阳　殷超　陶晓晨　张晓冬　李民伟　法尔科内·马利亚
李华　卡斯特罗·索菲娅　夏海青　李云　诺贝·马科斯　孙穆元
张德志　卫婷　林涛　曹玲妹　孟晗　栾海　陆峰　林彬　王宇石
赵旭千　苏仲洋　梅海斌　于春雨　范群立　方芳　风雪昆　于彦凯
林湧涛　阳舒华　宋岩　宋鹏　　（以入职先后为序）

图书在版编目（CIP）数据

万达商业规划 2015：持有类物业 / 万达商业规划研究院主编.
—北京：中国建筑工业出版社，2016.11
ISBN 978-7-112-20175-4

Ⅰ.①万… Ⅱ.①万… Ⅲ.①商业区—城市规划—中国 Ⅳ.① TU984.13

中国版本图书馆 CIP 数据核字 (2016) 第 304594 号

责任编辑：徐晓飞　张　明
执行编辑：刘易昆
美术编辑：陈　唯
英文翻译：喻蓉霞　王晓卉　郝　婧
责任校对：焦　乐　李欣慰

万达商业规划 2015：持有类物业
万达商业规划研究院　主编

*

中国建筑工业出版社出版、发行（北京海淀三里河路 9 号）
各地新华书店、建筑书店经销
北京雅昌艺术印刷有限公司制版
北京雅昌艺术印刷有限公司印刷

*

开本：787×1092毫米　1/8　印张：38　字数：950千字
2016年12月第一版　2016年12月第一次印刷
定价：**1000.00元**
ISBN 978-7-112-20175-4
（29644）

版权所有　翻印必究
如有印装质量问题，可寄本社退换
（邮政编码 100037）